Mathematics
MINUS Fear

First published in the United States and Great Britain in 2006 by
MARION BOYARS PUBLISHERS LTD
24 Lacy Road, London, SW15 1NL

www.marionboyars.co.uk

Distributed in Australia and New Zealand by Peribo Pty Ltd
58 Beaumont Road, Kuring-gai, NSW 2080

Printed in 2006
10 9 8 7 6 5 4 3 2 1

A CIP catalogue record for this book is available from the British Library.
A CIP catalog record for this book is available from the Library of Congress.

ISBN 0-7145-3115-4
13 DIGIT ISBN 9780-7145-3115-1

Set in Bembo 11pt
Printed in England by Cox & Wyman, Berkshire.

Mathematics MINUS Fear

by

Lawrence Potter

MARION BOYARS
LONDON · NEW YORK

I dedicate this book to Gina,
for being

so very, very small

CONTENTS

PART THREE: FEAR OF THE UNKNOWN

PART FOUR: CHANCE WOULD BE A FINE THING

INTRODUCTION

WHY?

School. Nobody forgets it. It lingers on in your head as a source of comedy and confusion for the rest of your life. It's the ritual of it. A weird tyranny of teachers. 'Stand up, sit down.' 'Hands out of your pockets.' 'Silence.' 'No skirts above your knees.' 'Put that gum in the bin.' Giant adult hands crashing down on desks. Red faces bellowing at you to 'Pay attention.' Tests. Examinations. Results. Reports.

At the centre of the system is the classroom. When a classroom is in use, it is bathed in a ghastly artificial light – just like the 'Engaged' sign for an aircraft toilet. Inside, rows of heads hunch over rows of orderly desks, upon which, every year, students stick gum, write their name, and unpack their books.

Each type of classroom has a special flavour. The geography classroom has maps of the world on its walls, and pictures of erupting volcanoes, whilst the English classroom has neat displays of 'Your Best Work'. The science classroom smells permanently of gas that has escaped from the taps for the Bunsen burners, whilst the gym just smells of sweat and pain.

But of all these flavours, it is often that of the mathematics classroom that has left the bitterest taste. Its walls are either left bare, or they are covered in uninspiring pictures of shapes and brightly-coloured multiplication tables. Inside this room waits the ordeal of quick-fire questions designed to expose and embarrass the weakest. In here, students open their little blue exercise books, and continue to fill it with row after row of calculations, whilst every furrowed brow and frustrated sigh signal the same unspoken message. 'I don't understand.'

The mathematics teacher is the remote and tyrannical figure who rules this little world. His intense love of numbers has damaged his interpersonal skills, and his strange physical features have somehow interfered with his love of humanity. He asks the questions, and dismisses hours of labour with a few scratchings of his red pen. And he scrawls incomprehensible explanations on the blackboard, and then expects his students to solve the problems on the sheet in front of them by some mystical form of osmosis.

Just like any environment, the conditions of the mathematics classroom affect its inhabitants. The fear of getting an answer wrong means that for most the best chance of survival is silence. This group of children always have their heads bowed over their books with furrowed brows. They never let their eyes wander around the room for fear of making eye contact with the teacher. They 'could contribute more to class discussion' (as their end-of-year report makes clear) – but they never will.

If you were one of the silent ones in the classroom, your voice swallowed by self-doubt, then this book is for you. I hope you will find use for it, even if you were one of those who were more intent on dazzling the spotty child at the front with reflected sunlight from your watch.

I offer this book as a kind of therapy for all those who found mathematics difficult at school. As part of the therapy, I need you to trust me to take you on a journey back through time to the trauma of that mathematics classroom. You will once again have to face the watery eyes of a mathematics teacher magnified horribly through the thick lenses of his glasses as he glares at you in expectancy. In what follows, this monster goes by the name of Mr Barton. And, as this man's gaze slowly eats away at your self-belief, you will once again have to suffer the disdainful glance of his prim sidekick, the-cleverest-girl-in-the-class. In the pages to come, she is known as Bernadette Pressman.

I am not saying that the process will be enjoyable, but it is necessary, if you wish to finally put the ghosts of Mr Barton and

Bernadette Pressman to rest. And you will not be alone. In the corner of the classroom sits Charlie Bissil. At heart, he is not a bad boy. But as lesson after lesson passes, he understands less and less of what Mr Barton says. There was a time when he stared earnestly at the board, and tried to construct some meaning from the symbols on it. But every time he felt close to success, Mr Barton and his red pen shattered his illusion. And so now he channels his energy into acts of small-scale rebellion. You will find him a useful ally in the ordeal you are about to face.

I hope that, in subjecting yourself to the therapy, you will come to understand some of those things that you never really understood at school, and which led you to believe that the whole subject of mathematics was a cruel joke. If this is the case, you will never again have to grope in your memory for rules copied down years ago into long-lost exercise books, nor will you have to continue to place your faith in the hands of Mr Barton (which were, after all, unnaturally sweaty). You will be able to face the world of mathematics with a renewed confidence in your own ability to understand it and to solve the problems that it throws at you.

And Mr Barton will cease to appear in your dreams to chase you through an eternal classroom, whilst assaulting you with questions from the nine-times-table and pieces of white-hot chalk.

PART ONE

NUMBERS IN YOUR HEAD, FIGURES ON PAPER

1 SMALL STEPS

It is the start of the school day. Most children are barely awake, but shuffle through the school corridors in a sleep-drugged state. In marked contrast to them, Bernadette is bright-eyed and straight-backed as she stands waiting for the beginning of the day's dose of education at the front of a disorderly line outside the mathematics classroom. Charlie, too, is beginning to show signs of life. He has already managed to get his football confiscated by accidentally kicking it into the windscreen of the headmistress's oncoming car. Now he is concentrating on navigating around the school building according to his own strict rules of motion.

Strangely enough, although one of the underlying principles of these rules is to minimise the distance that he travels, they often result in making him late for lessons, because he regularly has to stop and wait for a particular obstruction to clear. It is fine to push through a group of smaller children, but experience has taught him that the same tactic is less suitable for larger children or teachers. He has tried to explain this problem in the past when accused of 'wilful tardiness', but, in general, his teachers struggle to understand that it is part of his religion to always walk in straight lines.

He arrives just as the last of his classmates are entering the classroom for the first lesson of the day, and slips quietly to his place in the back corner, placing his backpack on the floor under the desk. Mr Barton is writing the day's date and a title on the board in spidery handwriting. Bernadette has arranged her pens and coloured pencils on the desk in front of her. They form a neat geometrical pattern around her eraser, which smells sweetly of strawberries. Charlie has forgotten his homework.

> 1. The postman comes every third day, the milkman comes every fourth day and the policeman comes every fifth day. One day they all turn up. How many days will it be before this happens again?

I know that you are probably a little defensive about your ability to 'do' numbers. However, first of all, I want to get your achievements into context. Forget about the pressure of the school mathematics test. Wipe from your mind that little destructive silence that your teacher left just after you announced that you had only scored 'five out of ten'. If you are able to do any sum – either in your head or on paper – that is a little miracle in itself. It means that you have already come a long way.

Let's see just how far. When you were born, you knew nothing. This is not an insult. No one knows much just after they have been born, except that it is a blessed relief to get out in the fresh air. As a result of interacting with the world, you gradually began to figure some things out. You knew that there is a difference between one aunt, two aunts, three aunts and four aunts, even if they were trying to distract you from figuring this out by making funny noises and invading your personal space. But any more than this number of aunts, and it was all too much. There could have been ten of them or fifteen of them – you wouldn't have known the difference.

And without any further help, that is as far as the human mind will get in Arithmetic. Occasionally, a child grows up without contact with other humans. Such a person is called a 'wild child'. If they are not discovered early in their life, this is the limit of their understanding of number. Once they have hit puberty, they are rarely capable of improving on this vague sense of the difference between one, two, three and four objects.

This is exactly the same stage as the wiliest of animals can reach. Some of the best natural mathematicians are certain species of birds, like the crow and the magpie. If you are one of those people who collect birds' eggs and put them in glass cases with neat labels, then please bear in mind this fact to avoid unnecessary cruelty to the poor mother. Don't raid a nest where there are four or less eggs, because the lady magpie will know that one of her future children has gone missing.

---

2. Eight prisoners usually serve their sentences in solitude. They are arranged in cells as the diagram below shows:

However, as they have been good, the jailer allows them to share cells, but they must arrange themselves so that there are four prisoners along each side of the prison. How many prisoners end up in each cell?

Leaving behind the animal kingdom, there have existed whole human civilisations that have not got further than this. Unlike the wild child, they all developed speech, but, in general, they only had numbers for 'one' and 'two'. They could deal with 'three' and 'four' by talking about 'two–one' and 'two–two', but beyond that the average Botocudan from the Brazilian rain forest would just point at his head, and look a bit sorry for himself. This is no comment on the intelligence of Botocudans. They were perfectly capable individuals. It's just that they had no need for numbers greater than this.

There is evidence of our inability to get beyond a concept of 'four' all over the place. For example, the Romans only gave normal names to the first four of their sons. Then the fifth son was always called Quintus ('the fifth'), the sixth son Sextus ('the sixth'), the seventh Septimus ('the seventh') and so on. Similarly, in the original Roman calendar (which only had ten months), the first four months had names unconnected to their position (Martius, Aprilis, Maius, Junius), but the rest of them were named from their order: Quintilis, Sextilis, September, October, November and December. Later, January and February were included, when it was realised that the months were falling out of step with the seasons, and Quintilis and Sextilis were renamed July and August after the emperors Julius Caesar and Augustus.

Now, this is not strictly relevant, but you are probably wondering how a member of the Botocudan tribe managed to keep track of things if she had no concept of a number greater than four. What happened if she found nine identical eggs in the nest of a macaw, and decided to carry them off home for breakfast? How would she know that she hadn't dropped one on the way back, if she didn't know the difference between nine and eight?

Well, the truth is that the Botocudans were perfectly capable of dealing with this sort of thing. They would take a tally. One way of doing this was that for each egg, they would pick up a pebble, or tie a knot in a piece of string, or cut a notch on a stick. And when they got home, as they took out each egg, they would throw away a pebble, or untie a knot, or cross out a notch. This way they could keep track of their belongings, and only ever really deal with the number 'one'. Each egg was a 'one' that they would record in whatever way was most handy.

In fact, one of the most common ways of keeping a tally was to use their own body. Each tribe would come up with a particular order for the different parts of the body. So the Botocudan woman would touch the little finger on her left hand when she put the first egg into her pouch, the second finger on her left hand for the second egg, and so on until she had used all her fingers for the first five eggs. For the sixth egg, she would touch her left wrist, for the seventh egg, her left elbow, for the eighth egg her left shoulder, and for the ninth egg her left breast. When she got home, she just had to go through this sequence again as she took out her eggs. If she ended up pointing at her left breast, then she hadn't dropped any.

This might all sound very primitive, but just to stop you from feeling smug, different methods of tallying have stuck with us through the ages. Up until 1828, the British exchequer sent out tax demands on tally sticks, and kept them as receipts in the basement of the Houses of Parliament. When the system was abolished in 1834, the politicians decided to burn all the sticks. Unfortunately, they lost control of the fire, and burnt down the Houses of Parliament by mistake.

So there you go. At the age of eighteen months, you are already at the same level as many civilisations ever reached altogether. As soon as your parents encourage you to count with your fingers (just like the Botocudans), and to give each new number a name (unlike the Botocudans), you have moved into a place that many inhabitants of this planet have never been. You are a little genius. And it won't be long before you can tell all nine of your irritating aunts to kindly leave you alone.

---

3. How can the numbers 1 to 9 be placed in these circles, so that each side of the triangle add ups to 20?

2 HOW MANY FINGERS?

Getting beyond the number four is only the beginning. As soon as you start counting, you bring into existence an infinite amount of numbers. And it is all very well to start giving special names to the first few of them, but you can't come up with new names forever, and even if you did, you wouldn't be able to remember them all. It's a bit like the Romans with their sons – after a while you give up trying to be original.

So the next challenge that you have to deal with is to understand the system that we use to cope with all these numbers. And the system that most people use is called the decimal, or base ten system. The best people to talk to about this are the Tibetans, because they stick to it the most completely. The Tibetans have come up with words for the numbers zero to nine (as we call them). They have also come up with words for every power of ten (as have we for the most part: ten, hundred, thousand and million). Then they can express in words any number they like by combining their words for zero to nine, with their words for the powers of ten. So, they would call the number 324: 'three-hundreds two-tens and four', or actually *'gsum-brgya gnyis-bcu rtsa bzhi'*.

Now you might be thinking, with a nationalist rush of anger, that the English language is every bit as logical as the Tibetan. But it isn't quite. For starters, in English, names for numbers get shortened to make them easier to say. 'Two-tens' becomes twenty. 'Five and ten' becomes 'fifteen'. Also, there is the mystery of 'eleven' and 'twelve'. They don't appear to have anything to do with 'two-and-ten' or 'one-and-ten', although one theory is that they are different – that is, don't contain any reference to ten – because they are so near to it in sequence. So 'eleven' derives from 'one left' (after ten) and 'twelve' from 'two left' (after ten). When we get to thirteen apparently we are getting too far away from ten to cope without being reminded of where we are.

And then there is the fact that we just got lazy when it came to making up names for powers of ten. While the words 'ten', 'hundred', 'thousand', 'million', 'billion' and even 'trillion' are all commonly used in English, the Tibetans went further. They also have a special name for 'ten thousand' and 'one hundred thousand'. We couldn't be bothered, which is a shame, because it would make writing cheques much easier. So we can't claim to be as logical as the Tibetans, but we can claim to be a lot more sensible than the Welsh. They came up with 'two-nines' instead of 'ten and eight'. Where is the sense in that?

> 4. On their daily rounds, three dustmen discover a pile of bins full of varying amounts of waste, which are blocking the public right of way and must be moved. It is important that nobody does more work than anyone else, and so they want to share the task equally. They calculate that there are sixty bins, of which twenty are completely full, twenty are half-full, and twenty contain no rubbish at all. How can they divide up the work, so that each person carries the same number of bins, and the same amount of rubbish?

You might have wondered why we count like this. And in doing so at such a tender age, you once again proved your potential for genius. That is exactly the same question that a fully-grown Aristotle asked, and he ranks as one of the greatest philosophers of all time: 'Why do all men, whether barbarians or Greeks, count up to ten and not to some other number?'[*]

The short answer to this question is: FINGERS. Your fingers are the most natural tool for counting that you have. At some point, people stopped using them as a tally (like the Botocudans), and started connecting them with numbers.

---

[*] Aristotle, *Problemata*, xv, 3 in Gow (1968) p1.

The long answer to the question is that in fact not everyone has counted like this. Although the vast majority of counting civilisations have used base ten, there are plenty of examples of people who used different bases. This may seem surprising. Our numbers and the way that we use them seem so natural that it is hard to believe that they do not just exist in the world that way. But the base ten system is just one of an infinite number of ways that we could have chosen to put numbers into a system. If you had eight fingers, rather than ten, for example, you would be using base eight, and be just as happy, except you would not be so good at playing the piano.

This is not just a hypothetical situation. Besides base ten, the most common number system used is the vigesimal system, or base twenty. Both the Maya and the Eskimos used base twenty, presumably because they were counting on their fingers and their toes – although what an Eskimo was doing without any shoes on, I don't know.

---

5. At Christmas you arrive home with a bag of presents. You have five young cousins, all keen to get their hands on the loot. The first of these takes half the presents and one more. You don't get far with the remaining presents before you meet the second cousin, who takes half of what you have left and one more. You stumble on, only to come up against the third cousin, who again takes half the presents and one more. The same thing happens with each of the two remaining cousins. Exhausted and dishevelled, you finally reach the sitting room, where you find your mother-in-law, who is waiting expectantly for her gift. You hand over the one and only parcel you have left. How many presents did you have to start with?

There are still elements of base twenty thinking around today. If you ask a mysterious stranger in the middle of a windswept English moor how far it is to the nearest pub, he might answer: 'Two score miles and ten'. He actually means 'Two lots of twenty miles and ten more' which to you and me is fifty. (The word 'score' for the number twenty has been used since Biblical times, when the average human lifespan was said to be 'three score years and ten' – i.e. seventy years. The word comes from the old method of keeping tally. When you got to the number twenty, you made an extra large cut, or score, in your counting stick.) Similarly if you ask a Frenchman for eighty onions, and in his surprise at the strength of your need for his national vegetable, he raises his eyebrows, and says: '*Quatre-vingt?*' What he means is: 'Four twenties?' Both of these people are using a base twenty system.

Just as base ten developed from counting the fingers on both hands, and base twenty from fingers and toes, a base five system developed in several civilisations through people counting on just one hand. To give you an idea of what you are missing, this is how a member of the Fulah tribe in West Africa would have dealt with numbers. It is a perfectly sensible way of going about things.

Firstly, he had special names for the numbers from one to four. To make life easier, let's say that these names were, in fact, 'one', 'two', 'three', and 'four'. He also had special names for the powers of five (5, 25, 125 etc.). Let's say they were as follows: 'five' (5), 'high-five' (25) and 'jackson-five' (125). He could then use this system to name any number he liked.

For example, take the number we call 'three-hundred-and-thirty-nine'. We have given this number its name because we think of it as being made up of three hundreds, three tens, and nine units. But the Fulah tribesman did not think of it as being made up in this way at all. He looked at it, and saw it as being made up of two jackson-fives, three high-fives, two fives and four units, and so he named it precisely that. You can check his thinking. He hasn't made any mistakes. It all adds up to 339. It is just a different way of looking at the same number.

(I should add here that I shouldn't really talk about the number 339 as if that is the only way of representing this number in symbols. It isn't, and the Fulah tribesman, if he had got around to writing numbers using symbols, would not have written it like this. But that is something that I will come to later. For now, when I write 339, I simply mean the number that we are referring to when we write down these symbols.)

It is possible that your mathematics teacher never told you about all of this. It is possible that he kept it to himself, tucking away his knowledge in his tattered leather briefcase next to his Tupperware box containing corned beef sandwiches and an overripe tangerine. But it is all true. The way we count is the result of the design of our bodies. It is a system that we have made up to deal with the consequences of inventing number. And it is by no means an easy one to understand.

---

6. i. What would we call this Fulah number:
   'four Jackson-fives, three high-fives, two fives and one'?
   ii. What would a Fulah call our number:
   'four-hundred-and-seventy-three'?

---

3 OUTSIDE THE SUPERMARKET

For most of my time at primary school, doing mathematics was the same as doing sums. You had two choices when faced with a sum. You could either do it in your head, or you could do it on paper. Doing it in your head was harder, and you got more respect. If you could do a sum very quickly, the chances were that some of your classmates would gasp, and mutter: 'He's CLEVER'. Of course, this only lasted until the beginning of your first year at secondary school. After that, if you did a sum in your head quickly, you were more likely to be beaten up.

Still, there is no denying that doing mathematics in your head is a useful skill, if only to make sure that you never get short-changed. I once had a very short-lived stint as a barman. I was expected to add up the price of a round of drinks in my head. I found this very hard, especially on a Friday night, when the pub was packed, and someone had already ridiculed me for being the only barman not wearing a white shirt (mine was blue – I still don't know what the problem was). I began to make more and more mistakes, and my confidence ebbed away. It began to affect my other duties as a barman. I lost my ability to pour a decent pint of lager. My hands started to tremble. The punters became more aggressive. My boss had a go at me for taking too long to deal with an order, and for failing to add up the prices correctly.

At last, one customer made the simple request of a pint of soda and lime. Just one pint. No problems totting up the price of this order (I thought), and no problems in pouring it. It was the moment of calm that I needed to collect my thoughts. I poured in the lime, and then pressed the button to squirt the soda into the glass from the siphon. As I turned to the customer with an air of nonchalance to tell him the price of his drink, I let go of the button. It was stuck. The siphon continued to shoot out soda water in an impressive jet. I tried to stop it by covering its nozzle with my finger, but this only had the effect of increasing the power of the stream of soda, and sending it fizzing all over the bar and several of the customers. As the whole pub turned to stare, I found myself grappling with the siphon and its metallic cord as if with a futuristic serpent. I had finally wrestled it to the ground, and was about to bite off its head, when my boss calmly turned it off at the main pump. The siphon went limp, and I was taken off bar duty. I spent the rest of the evening arranging crisps in boxes so that the customers could see what flavour they were.

You might wonder what this story has to do with anything. But remember that the root cause of this disaster was my inability to do sums in my head. This flaw in my mental make-up gradually undermined my confidence, and I hold it totally responsible for leaving me lying in a fizzy puddle on the dirty floor of a backstreet pub.

---

7. The letters A to G each stand for one of the numbers 1, 3, 4, 5, 6, 8 and 9. Can you work out which letter stands for which number from the following facts:

$A + A = B$	$A \times A = DF$	$A + C = DE$
$C + C = DB$	$C \times C = BD$	$A \times C = EF$

In order to improve my mental arithmetic, I went out into the world to find out the different techniques that people use to deal with numbers in real life. Or rather, I stood outside a supermarket in South London, and asked people mental arithmetic questions. It was only one afternoon – just a few short hours – but it taught me many things.

Firstly, I learnt that the world is a hostile place, and people on their way to supermarkets are mean. Many people just pretended that I didn't exist. Others marched past me with a cruel smile playing on their lips. Secondly, I learnt that women are much nicer than men, although it is best to avoid mothers with more than three children in tow. Thirdly, and most importantly, nothing has changed since school. No one has grown up. Everything is exactly the same.

I wonder if you recognise any of these people from your days in a classroom. Do you remember a boy who, after scoring ten out of ten in a mental arithmetic test, pumped his fist and yelled 'I am the King'? In fact, so confident was he in is own ability that he demanded that the teacher give him an extra-hard multiplication sum, just to underline his talent. Or do you recall the girl who just got redder and redder and redder as each question was asked, until she burst into a fit of hysterical laughter, and had to leave the room? What about the boy with an opinion of himself that was slightly too high, who called out the wrong answer to a question that he was not being asked? And the girl who was being asked, and who was trying her level best, who then flew into a rage at him, and questioned his manhood? Ring any bells? Well, they were all there, outside the supermarket.

And not only were the characters the same; their attitudes towards doing sums were the same as well. They were obsessed with knowing whether they had got the answer right. Some of them even asked me to give them a mark at the end of the survey. One man came back half-an-hour later to tell me that he had worked out the answer to question nine. He claimed it was seven – he was wrong.

Many of them were nervous about having to do mathematics in the open. Several people told me that they were rubbish at arithmetic at school. They looked fearfully around them as they answered the questions. And one woman really did run away when I asked her to do a division in her head. Presumably, I had reopened an old and painful wound, and I sincerely apologise to her.

4   PUTTING TWO AND TWO TOGETHER

Due to my afternoon spent next to a spectacularly long line of trolleys, I can tell you that most people do addition and subtraction sums in their heads just like they would do them on a piece of paper. When they worked out 76 + 22, they added the 2 and the 6 to get 8, then the 7 and the 2 to get 9, and they gave the answer correctly as 98. Similarly, they gave the answer to 76 − 24 to be 52, by subtracting the 4 from the 6 and then the 2 from the 7.

However, things get a bit harder to deal with mentally when you start having to do some 'carrying' or 'borrowing' − just as in real life it is easier to walk up some stairs empty-handed, and to pay for a car with your own money. So, people had more problems with the sums 59 + 64 and 74 − 29. Some stuck to doing the paper-and-pencil method in their heads. But these people were more likely to get the answer wrong, or to be unsure whether they had got the correct answer. And I think that this is because if you work in this way, you are not really thinking about the actual number (e.g. 59), but are splitting it into two parts (e.g. a 5 and a 9) which are not obviously related to the original number. A 5 and a 9 make 14, not 59. If you see what I mean.

Anyway, other people came up with different ways of going about these problems that I thought you might find helpful. The most common thing to do is to break up the second number into bits and do the addition and subtraction in stages. So for 59 + 64, you can split the 64 into a 60 and a 4, and do the sum as follows:

$59 + 60 = 119$ and then
$119 + 4 = 123$

or, $59 + 4 = 63$ and then
$63 + 60 = 123.$

Or, since $59 + 64$ is the same as $64 + 59$, you can split the 59 into a 50 and a 9, and solve the problem like this:

$64 + 50 = 114$ and then
$114 + 9 = 123$

or, $64 + 9 = 73$ and then
$73 + 50 = 123.$

And similarly for the subtraction $74 - 29$, you can break the 29 up into a 20 and a 9. So:

$74 - 20 = 54$ and then
$54 - 9 = 45$

or, $74 - 9 = 65$ and then
$65 - 20 = 45.$

In all of these examples, each of the stages is easier to do than the original sum.

---

8. Four children are involved in an Easter-egg hunt. When the hunt is over, the four children have collected forty-five chocolates in total, but they each have different amounts. If the first child had two more, the second child had two less, the third child had twice as many, and the fourth child had half as many, they would all have the same number of chocolates. How many chocolates does each child have?

Another thing to do is use the same sort of idea in a different way. Nobody likes dealing with sevens, eights and nines. (Remember that most civilisations didn't even know that they existed.) People (and magpies) much prefer dealing with ones, twos and threes. So play to your strengths. Add (or subtract) whatever you need to get up to (or down to) a multiple of ten, and often you will make the sum much easier.

So, for 59 + 64:
first add 1 to get 60   (59 + 1 = 60)
then add the remaining 63   (60 + 63 = 123)

and, for 38 + 23:
first add 2 to get 40   (38 + 2 = 40)
then add the remaining 21   (40 + 21 = 61)

and for 74 − 29:
first take away 4 to get 70   (74 − 4 = 70)
then take away the remaining 25   (70 − 25 = 45).

Or, especially when dealing with nines and eights, ignore them altogether.

For 59 + 64:
work out 60 + 64   (60 + 64 = 124)
then take away 1   (124 − 1 = 123)

for 38 + 23:
work out 40 + 23   (40 + 23 = 63)
then take away 2   (63 − 2 = 61)

for 74 − 29:
work out 74 − 30   (74 − 30 = 44)
then add 1   (44 + 1 = 45).

For those of you who are making snorting noises, and patting yourselves on the backs because you thought all the sums I have mentioned are too easy, please remember two things. Firstly, by doing any sums at all, you are demonstrating your mastery of counting and the base ten system of numbers, which would have put you in the elite of any society you cared to mention up to about five hundred years ago. And secondly, sums can always get harder. They always seemed to do exactly that when teachers asked you to turn to the next page in your textbook. Now why couldn't they have come out with one that got easier?

9.a) You go fishing for the day. In the morning, you use worms and catch twenty-eight fish. In the afternoon, you use dynamite, and catch seventy-six fish. How many fish do you catch in total?

b) You are standing in a smoke-filled pub, frowning at a darts board. Your girlfriend has never played before, and has scored seventy-four points so far. You have always thought yourself a talented amateur, but have only scored forty-eight points. How many points do you need to catch up?

---

5 GO FORTH AND MULTIPLY

So, you've mastered adding and subtracting in your head. But I'm afraid it all gets more troublesome when it comes to multiplication. For centuries people have been trying to avoid it. The Babylonians and ancient Egyptians got their most intelligent scholars to write down giant times tables, so that they wouldn't have to bother doing it at all. And many other cultures have come up with clever ways of avoiding the issue.

Try this. Pretend you don't know the answer to the sum 7 × 9. Now, hold your hands out in front of you, palms up, and fingers flexed. On one hand, I want you to close as many fingers as the first number of the sum is greater than five (which in this case is two), and on the other hand I want you to close as many figures as the second number of the sum is greater than five (which in this case is four). Now, I want you to multiply the total number of closed fingers on both hands by ten, and also multiply the number of stretched fingers on the first hand by the number of stretched fingers on the second hand. Then, add together your two results. You should find that you get the answer to the original problem (sixty-three). This is a method that good honest folk from all over the world have used at various times over the centuries, although, unless you are a mutant, it breaks down as soon as you have to multiply two numbers greater than ten, because you run out of fingers to bend.

No matter how many clever tricks you work out, you can't run from multiplication forever. So it's best to get some unpleasant truths out of the way. First of all, you have to learn your tables.

It's not me telling you this. Nicholas Chuquet, one of the very first men to reintroduce arithmetic to Western Europe after the Dark Ages, was already advising the very same thing in 1484 in his book *Triparty en la science des nombres*.

I thought I could get by without them. I remember being very smug during my two-times-table test, as the teacher read out the questions. '7 × 2', he said, and I chuckled merrily, whilst simply adding on a few twos to my answer for 4 × 2. 'It's just fast adding,' I said to myself.

That is what I thought. And I still thought it when I aced my three-times-tables test, AND the four-times-tables test, AND the five-times-tables test. But I began to run into problems during the six-times-tables test, and the seven-times-tables test was a total disaster. I did so badly, I had to retake it the following day during lunchtime, with my teacher asking the questions whilst chewing on a horrible mixture of tuna and mayonnaise.

---

10. A farmer wants to make chain of thirty links out of the six pieces of chain he already has. Each piece of chain is five links in length. It costs eight cents to cut open a link, and eighteen to weld it back together again. A brand new thirty-link chain costs a dollar and a half. What is the cheapest way for him to get his chain and how much does it save him?

I am not saying that you have to know every single result from your times-tables. You just have to have most of them in your heads, and you can work out the rest from there. Most people are pretty happy with the two-times-table up to the five-times-table. That doesn't really leave many new number facts to learn. For example, 3 × 8 is the same as 8 × 3, whilst 4 × 9 is just the same as 9 × 4. And, if you run into trouble, you can always use a bit of adding to get you to the answer. The majority of people I talked to outside the supermarket didn't know 7 × 8 straight off. They worked it out from number facts that they did know. For some reason, a large number of them knew what 7 × 7 was, and then they just added one seven to get the answer.

Once you know the times-tables, it is possible to break down harder multiplications into doable steps. The most common method is to break up one of the numbers involved.

For example:

16×4 ('16 lots of 4') is the same as
$(10 \times 4) + (6 \times 4)$ ('10 lots of 4' plus '6 lots of 4').

So $16 \times 4 = (10 \times 4) + (6 \times 4) = 40 + 24 = 64$.

The initial sum lies outside the limit of normal times-tables, but by breaking it down into two multiplications, you can make use of number facts that you know, or can quickly work out.

Similarly:

28×6 ('28 lots of 6') is the same as
$(20 \times 6) + (8 \times 6)$ ('20 lots of 6' plus '8 lots of 6').

So $28 \times 6 = (20 \times 6) + (8 \times 6) = 120 + 48 = 168$.

It is often a good idea to switch the multiplication around.

5×24 ('5 lots of 24') is the same as 24×5 ('24 lots of 5').

Then:

24×5 ('24 lots of 5')
is the same as
$(20 \times 5) + (4 \times 5)$ ('20 lots of 5' plus '4 lots of 5').

So $5 \times 24 = 24 \times 5 = (20 \times 5) + (4 \times 5) = 100 + 20 = 120$.

(For some reason, I find it much more obvious that 'twenty-four fives' is the same as 'twenty fives plus four more fives', than that 'five twenty-fours' is the same as 'five twenties plus five fours'. So switching the sum round makes it feel more comfortable to me.)

This 'splitting' method can be applied to harder multiplications in a (sometimes vain) attempt to reduce it to something easier to deal with mentally. For example:

17 × 13 ('seventeen thirteens') becomes
(10 × 13) + (7 × 13) ('ten thirteens plus seven thirteens').

Or, if you switch round the sum, 17 × 13 is the same as 13 × 17. And then:

13 × 17 ('thirteen seventeens') becomes
(10 × 17) + (3 × 17) ('ten seventeens plus three seventeens').

But now the numbers are beginning to get a bit large.

---

11. A packet of Pringles contains forty-three crisps. You own five packets. You open the first and start eating. You have no memory of the following thirty minutes, but at the end of this period of time, you find yourself sitting on the floor surrounded by empty packets and crumbs. How many Pringles have you eaten?

There are other methods that can help with particular problems. For example, instead of working out 19 × 4, you can work out 20 × 4, and then take one four away from your answer:

20 × 4 = 80.
So 19 × 4 = 80 − 4 = 76.

Similarly, instead of figuring out 48 × 6, it is easier to work out 50 × 6, and then take away two sixes to get your answer.

$50 \times 6 = 300$.
So $48 \times 6 = 300 - 12 = 288$.

Another useful technique is to use doubling, which most people find a fairly natural process.

For example, 4×18 becomes 'double 18, and double again':

$4 \times 18 = 2 \times (2 \times 18) = 2 \times 36 = 72$.

4×24 becomes 'double 24, and then double again:

$4 \times 24 = 2 \times (2 \times 24) = 2 \times 48 = 96$

And 8×16 becomes 'double 16, double it again, and double it once more':

$8 \times 16 = 2 \times [2 \times (2 \times 16)] = 2 \times (2 \times 32) = 2 \times 64 = 128$.

Before I abandon multiplication and leave it for dead, I would just like to say one thing. 17×13 is not the same as $(10 \times 10) + (7 \times 3)$. It looks like it might be – but it isn't. The best way of showing this is to pretend that you want to find out the area of a field that is 17 metres long and 13 metres wide. To work out the area of this field quickly, you would multiply 17×13 (in your head – using one of the techniques above). But if you wanted to complicate the issue, you could divide the field up into four sections, and work out the area of each one, like this:

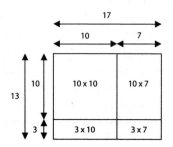

From the diagram you can see that 17 × 13 is not the same as (10 × 10) + (3 × 7). This is only part of what it is. It turns out that 17 × 13 is in fact the same as: (10 × 10) + (3 × 7) + (10 × 3) + (10 × 7).

> 12. A man called Sissa ben Dahir is said to have been the inventor of the game of Chess. The Indian King Shirham was very impressed by it, and asked Sissa what he wanted as a reward. Sissa said that all he wanted was a piece of grain to place on the first square of the chessboard, two pieces to place on the second square, four pieces for the third square, eight pieces for the fourth square, and so on until each square on the board had been filled. The king was amazed at his lack of ambition. But how much grain did Sissa ask for?

Ummm…division. Division is nobody's friend. So I am not going to dwell on it for too long. Except to say that it is just backwards multiplication (amongst other things). So 63 ÷ 9 is the same question as: 'What do you multiply nine by to get sixty-three?' 72 ÷ 8 can be thought of as: 'What do you multiply eight by to get seventy-two?' 132 ÷ 12 is equivalent to: 'What multiplies twelve to give one-hundred-and-thirty-two?' And 189 ÷ 9 translates as: 'How many nines in one-hundred-and-eighty-nine?' By thinking of such a problem in this way, you can escape having to worry about division completely.

Also, because division is connected to multiplication, you can use the same sort of tricks.

So, you can split the number to be divided:

$228 \div 4$ is the same as $(200 \div 4 + 28 \div 4)$
$228 \div 4 = (200 \div 4) + (28 \div 4) = 50 + 7 = 57$.

$642 \div 6$ is the same as $(600 \div 6 + 42 \div 6)$
$642 \div 6 = (600 \div 6) + (42 \div 6) = 100 + 7 = 107$.

Or you can half and half again:

$228 \div 4$ is 'half of 228, and half again'
$228 \div 4 = (228 \div 2) \div 2 = 114 \div 2 = 57$.

$184 \div 8$ is 'half of 184, and half again, and half again'
$184 \div 8 = [(184 \div 2) \div 2] \div 2 = (92 \div 2) \div 2 = 46 \div 2 = 23$.

Or you can do the division in two stages. To work out $216 \div 12$, you can first divide 216 by 2, and then divide that answer by 6:

$216 \div 12 = (216 \div 2) \div 6 = 108 \div 6 = 18$

And to work out $336 \div 8$, you can first divide 336 by 2, and then divide that answer by 4:

$336 \div 8 = (336 \div 2) \div 4 = 168 \div 4 = 42$

But in general, division is best tackled using pen and paper.

6 'COUNTDOWN'

So, not only have you learnt how to count, but you have also figured out how to deal with our base ten system of numbers. You could have stopped there, but you went on to master how to add, subtract, multiply and divide. If you had lived in the Mediaeval Ages, this would have put you in the academic elite. The average man would have looked at you as some kind of magician, and been tempted to burn you at the stake for your wizardry.

But in the modern world, if you want this kind of recognition, it's not good enough just to do well in arithmetic tests at school. You have to take people on in mental combat: on 'Countdown.'*

There is one particular section of the show called 'The Numbers Game'. In this section the contestants are given six numbers which they have to combine by addition, subtraction, multiplication and division in order to get as close as possible to a randomly-selected three-figure number.

The numbers are all either single-digit numbers, or 25, 50, 75 and 100. Not all the numbers have to be used, and the contestants are not allowed to combine two of the numbers to make a third number by just writing the digits next to each other (e.g. 3 and 7 cannot be put together to make 37). For example, if the contestants were given the numbers 25, 6,

---

* For those of you who are unaware of the existence of this cultural phenomenon, it is a TV quiz show broadcast on UK daytime TV, much loved by students and pensioners.

2, 3, 8 and 7 to make the number 137, they could multiply the 25 by the 6, and then subtract the 2, the 3 and the 8. It is possible to specify how many of each type of number you want, and, in general, it is best to choose five single-digit numbers and one other. Any other selection makes the challenge much harder.

There are a couple of general strategies that give you a good chance of solving each individual problem. Let's say that the following numbers are placed on the board: 4, 6, 9, 3, 7, and 50, and that your target number is 412.

The first thing to do is to look for the nearest multiple of 50 to 412. In this case, it is 400. Then you need to think about how many 50s make 400, and how much you have got left to get to 412. In this case, eight 50s make 400, and you have got 12 left to get to the target. In other words, $412 = (8 \times 50) + 12$.

So you need to make 8 and 12 by combining the little numbers. You can multiply 3×4 to get the 12. All that remains is to make 8 from the 9, 6 and 7. This is not so easy, but it is possible: $9 - (7 - 6) = 8$.

Therefore, you hastily scribble down on your piece of paper:

$$9 - (7 - 6) = 8 \text{ and } 50 \times 8 = 400$$
$$3 \times 4 = 12 \text{ and } 400 + 12 = 412.$$

Alternatively, you can try and work from the multiple of 50 on the 'other side' of the target number, which, in this case, is 450. You need nine 50s to get to 450, and then you need to take away 38 to get to the target: $412 = (9 \times 50) - 38$.

Therefore, you need to make 9 and 38 using the little numbers. 9 is one of the little numbers given to you, and 38 is $(7 \times 6) - 4 = 38$. As the final seconds of the round tick away, you sit back in your chair and relax, because you know the points are in the bag:

$$9 \times 50 = 450$$
$$(7 \times 6) - 4 = 38 \text{ and } 450 - 38 = 412$$

Sadly, this technique does not always work. In general, it runs into trouble when the nearest multiple of the large number to the target number is too far away for the remaining difference to be made using the small numbers. In such cases, it is necessary to be a little more cunning.

1, 2, 3, 4, 8 and 75 are up on the board. The target number is 633. The nearest multiple of 75 to 633 is 600, which is eight 75s. There is an 8 amongst the small numbers, but if you use it, it is not at all obvious how to make the remaining 33 from the small numbers that are left.

In this case, you must play with the first multiplication to get its answer closer to the target. $8 \times 75 = 600$. You are 33 away. Now, it is necessary to see if it is possible to multiply 8 by the other smaller numbers and get an answer that is close to 33. In this case, 8×4 is 32.

You can use this result as follows:

$$8 \times (75 + 4) = (8 \times 75) + (8 \times 4) = 600 + 32 = 632.$$

Now you are much closer to the target number than after the initial multiplication. All that remains is to add on the 1 to get to 633. Your answer is:

$$8 \times (75 + 4) = 632 \text{ and } 632 + 1 = 633.$$

One more. This time the numbers are 1, 2, 3, 6, 9 and 100, and the target is 448. The nearest multiple of 100 to 448 is 400. Four 100s make 400, and there is 48 left. It is possible to make the 4 from the 1 and the 3, but not easy to see how to make the remaining 48 from the numbers that are left.

So, you go back to the first multiplication: $4 \times 100 = 400$. You are 48 away from the target. You want to multiply the 4 by the small numbers that are left to get as close to 48 as possible. The best you can do is $4 \times 9 = 36$.

You can use this result as follows:

$4 \times (100 + 9) = (4 \times 100) + (4 \times 9) = 400 + 36 = 436.$

You are now only 12 away from the target, and you can make 12 using the 2 and the 6. So the answer is:

$1 + 3 = 4$ and $4 \times (100 + 9) = 436.$
$2 \times 6 = 12$ and $436 + 12 = 448.$

Just to show how flexible this method is, it is possible to solve this problem in an alternative way by approaching the target from 500, the multiple of 100 on the 'other side' of 448.

$5 \times 100 = 500$, and it is possible to make the five from the 2 and the 3. You are now 55 away from the target. You want to multiply the 5 by the small numbers that are left to get as close to 55 as possible. The best you can do is $5 \times 9 = 45.$

Then: $5 \times (100 - 9) = (5 \times 100) - (5 \times 9) = 500 - 45 = 455.$

You are now just 7 away from the target, and this 7 can be made from the 1 and the 6. You can solve the problem as follows:

$2 + 3 = 5$ and $5 \times (100 - 9) = 455$
$455 - 1 - 6 = 448.$

---

13. See if you can now demonstrate your mastery of 'The Numbers Game' by solving these problems:

a) The numbers are 1, 2, 3, 6, 8 and 25. The target is 213.
b) The numbers are 1, 2, 3, 5, 9 and 75. The target is 377.
c) The numbers are 1, 4, 6, 7, 9 and 100. The target is 648.
d) The numbers are 2, 3, 4, 8, 9 and 75. The target is 862.

---

7 PUTTING NUMBERS TO PAPER

I think we should take a moment and look back over all your achievements. You have tackled mental arithmetic, but is that enough? What about the other side of the coin? What about all those sums that it is not possible to do in your head – those sums that have to be written down? Can you remember how to tackle the monsters of long multiplication and long division? Because if you can, you are in the minority.

Before we plunge nose-deep into such muddy waters, it is worth looking at where these monsters came from. The history of written arithmetic is long and complicated, and, as is already clear from the previous chapter, it took thousands of years for societies to develop to the point where they could tackle complex number problems. Nobody should ever suggest that it is easy.

When Christ was born, the most common method of recording numbers was the Greek system. The Greeks took their alphabet, and assigned a letter to each number from one to nine, to each ten from ten to ninety, and to each hundred from one hundred to nine hundred. Sadly, they ran out of letters in their alphabet, so they brought an old letter back, and borrowed a couple from the Phoenicians to make up the numbers.

So, using their system, α was one, β was two, θ was nine, ι was ten, κ was twenty, ρ was one-hundred, and σ was two-hundred – just to name a few examples. By combining the letters, they could produce any number up to 999. For example, they would write 112 as $\rho\iota\beta$, 229 as $\sigma\kappa\theta$ and 354 as $\tau\nu o$. They were perfectly capable of dealing with larger numbers. They invented symbols for a thousand (/) and ten thousand (M). If they wanted to express two thousand, they wrote / β, and if they wanted to express twenty thousand, they wrote M with β above it:

$$\overset{\beta}{\mathrm{M}}.$$

This probably sounds very sensible, but unfortunately (or fortunately – depending on how you see it) it makes calculations very difficult. We only have to deal with combinations of ten different symbols, where the Greeks had to deal with twenty-seven. For us, adding twenty and thirty is essentially the same as adding two and three, but this is not true for the Greek system. There is no connection between the symbol for two and the symbol for twenty. As a result, the Greeks tended to use abaci to do their sums, rather than use written methods.

In fact, they were quite snooty about calculations in general. Philosophers and mathematicians were much more interested in number patterns than working out sums, because they felt that such patterns were behind the very fabric of the universe. The Greeks got so excited by this sort of thing that they became downright religious about numbers.

The Pythagoreans were the followers of Pythagoras, the ancient Greek philosopher and mathematician who is widely thought of as discovering Pythagoras' Theorem ($a^2 + b^2 = c^2$). This is not strictly true as there is evidence that the Egyptians were using it to build pyramids and the Chinese were using it to survey land a long time before he came on the scene – but he and his followers certainly made use of it, and helped introduce it to the Western world.

According to the Pythagoreans, the numbers one to four had special properties relating to the shapes that make up the universe. One was the point, two the line, three the surface, and four the solid.

Ten was the 'perfect number' because it is the sum of one, two, three and four – the point, the line, the surface and the solid.

Some numbers were thought to be good. Some numbers were thought to be bad. Odd numbers were considered to be male, because if you arrange an odd number of dots in pairs there will be one dot left at the end sticking out. Even numbers were female, because if you arrange the dots there are no projecting dots.

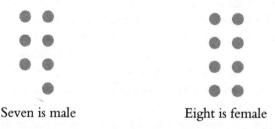

Seven is male   Eight is female

For people of this sort of bent, the alphabet system came in quite handy, because it meant you could work out a number-value for a word or a name, and thus draw conclusions. This idea is called numerology. It has been, and still is, popular among many cultures. It follows on from the mystical belief that numbers somehow capture truth in a special way.

You might think that this sounds very dubious, but plenty of people still hold such beliefs about numbers. We generally consider thirteen to be unlucky, and seven to be lucky. The Japanese don't like the number four, because their word for it sounds like their word for 'death'. And they are not too keen on the number nine either, because it sounds like the word for 'pain'. As a result, hotels often avoid having any rooms labelled with a number that contains a four and a nine, and the launch of the new Renault 4 was a complete disaster – people just didn't feel comfortable sitting in the front seat of a Renault 'Death'.

You can prove just about anything with numerology, if you work at it hard enough. One of the most popular pastimes in this area is to work out who (or what) is the Beast of the Apocalypse. It is prophesied to turn up just before the End of Days to spread evil and terror among mankind. Clearly, if it turns up to your house for a key party, it is important that you will be able to spot it, so John the Apostle helpfully mentioned that it will have a sign on it. This sign is the number 666.

Now, some people think that this sign will be camouflaged in some way, because the Beast might want to spread some undercover terror. And they reckon that numerology might be the key to blowing the whistle. However, this creates a bit of doubt, because everything depends on how you decide to decipher the sign. Over the years, various people have accused all sorts of other people of being the Beast, and mostly just because they didn't see eye to eye. The Christians claimed Emperor Diocletian was the Beast because the letters in his Greek name added up to 666, and because he persecuted them. The Catholics claimed Martin Luther was the Beast because the letters of his name in Latin added up to 666, and because he set up a rival Church. The Seventh Day Adventists claimed that the Pope was the Beast, because the letters of one of his titles added up to 666, and because they weren't keen on his religious views. More recently, other candidates have included Hitler, Prince Charles, Bill Gates, Viagra and George Bush Jnr.

Numbers also play a big role in the ancient Chinese art of Feng Shui. One of the main tools that practitioners of Feng Shui use is called the Lo Shu magic square. The square comes from a Chinese legend which goes back as far as the 21st century BC. The Emperor Yu was walking along the banks of the river Lo when he met a divine tortoise. This tortoise had magical markings on its back: the Chinese symbols for the numbers one to nine inclusive ordered in a 3 × 3 grid. From the markings on the tortoise's back, Emperor Yu was able to work out the magic number of fifteen, since this is the sum of every row, column and diagonal.

4	9	2
3	5	7
8	1	6

In Feng Shui, the magic square is applied to a building, and each of the numbers signifies an area of life. So using the above positioning of the numbers, we can say that the south part of the building (which is governed by 8, 1 and 6) is the area for prosperity (8), good luck (1), and power, respect and fame (6). This information would clearly be very useful. For a start, you would know exactly where to sit when you are next plotting to take over the world.

For the Chinese, numbers (like everything else) have yin and yang qualities. Even numbers are yin and odd numbers are yang. Generally speaking yang numbers are more desirable than yin ones. Eight is the most auspicious number to the Chinese because it is the 'yinnest' of the yin numbers, or in other words, the most unlucky. The thinking behind this is that when you are at rock bottom the only way is up, so eight represents a potential change in fortune.

As we have seen, Greek numbers were excellent if you wanted to identify your neighbour as the Antichrist, but not so useful if you wanted to work out the price of beans. With the rise of the Roman Empire, Roman numerals gradually took their place. The Romans had symbols for one (I), five (V), ten (X), fifty (L), one hundred (C), five hundred (D) and one thousand (M). These were much more useful for addition and subtraction, but they were very unwieldy for more complicated sums. And it took a long time to write large numbers. For example, when the BBC wanted to represent the year 1988 on its title page, this is what it had to write: MCMLXXXVIII. Even so, Roman numerals stuck around for a long time. They were still used in some mathematics books in the 16th century, and they only really disappeared when printed books became widespread. Meanwhile, since they were so cumbersome for complex calculations, merchants stuck to using the abacus and waited for something better to turn up.

---

14. Hoping to keep her class quiet for a lesson, a primary school teacher once asked her class to add up all the numbers from 1 to 100. Unfortunately, one of the children was Carl Friedrich Gauss, who was destined to become one of the greatest mathematicians of all time. He found a quick way of solving the problem, and put his hand up to give the correct answer after a couple of minutes. The teacher grudgingly gave him a gold star, and got through the rest of the lesson by playing 'numbers bingo'. How did he do it?

Whilst the Europeans were wallowing in the Dark Ages, other parts of the world were making impressive advances in their number systems. It all started in India, when, in about the 7th century, the Hindus came up with a decimal place-value system. They realised that they could write numbers far more economically, if the position of a digit influenced its value. So, they invented the 'units column', 'the tens column', 'the hundreds column' and so on. (I will forgive you if you are suffering flashbacks to your primary school classroom.)

As a result of this invention, the Hindus could represent any number they liked, using just nine symbols. So, where the BBC took eleven symbols to write just one number, the Hindus only took four: 1988. The one represents 'one thousand', the nine represents 'nine hundred', the first eight represents 'eight tens', and the second eight represents 'eight units'.

However, it wasn't all plain sailing for the Hindus with their new invention. Their main problem was how to show that a column was empty. Initially, they just left a gap, but this caused trouble, because it was very difficult to say whether a space in a number represented one gap, or two gaps, or three gaps. And yet the size of the space had a big influence on the size of the number. For example, 2 space 3, could be two-hundred-and-three, two-thousand and-three, or twenty-thousand-and-three, depending on how many gaps you thought it represented. So finally, at some time during the 9th century, they came up with the symbol '0' to represent an empty column.

From India, the decimal place-value system spread to the Middle East. Copies still exist of a book by a man called al-Khwarizmi (approximately 790-840) with the title of *A Book on Addition and Subtraction after the method of the Indians*. In this book, al-Khwarizmi gives methods for addition, subtraction, multiplication and division. Later Arab mathematicians even began to experiment with extending the decimal place-value system to deal with fractions.

The new numbers finally made their way to Europe around the 14th century. The Italian mathematician Fibonacci travelled with his merchant father to North Africa where he learnt about the Arabic number system. In 1202 he published a book, *Liber Abaci*, which showed the possible practical applications of the new methods. The book was well received by European academics, but the use of the decimal number system didn't become widespread until after the invention of the printing press in the 1400s.

You would have thought that the long-suffering merchants there would have been grateful for a system that made calculations quicker and easier. But, on the contrary, many refused to use them, claiming that it was a satanic practice, and that the numbers themselves contained dark magic. Another disadvantage for the new numbers was that their very simplicity made forgery a threat to their use in trade. To solve the problem, enterprising merchants came up with the idea of cheques.

So, it took a long time for societies to create a written number system that would allow them to deal easily with written calculations, and we, in the West, were particularly slow to catch on. Just to rub it in, several other civilizations have invented a place-value system over the years. The Babylonians figured it out in the 19th century BC, although they worked with a number system that was a mixture of base ten and base sixty. They had a symbol for ten, which looked a bit like < and a symbol for one, which looked like Y.

To write any number up to sixty, they just combined the correct number of each symbol. So forty-three was ⚡𝗬𝗬𝗬 and 55 was ⚡⚡𝗬𝗬 . Beyond this they used a place-value system, but because they worked in base sixty, the values of their columns were different. Each successive column had a value sixty times greater than the previous one. So, the first one was units, but the second was for 60s, the third for 3600s, and so on.

Therefore, to write the number that we would call 581, the Babylonians would not consider it, like us, as made up of units, tens and hundreds (in this case five hundreds, eight tens and one unit). They would see it as a combination of units and sixties. For them, 581 was nine 60s and 41 units $[(9 \times 60) + (41 \times 1) = 540 + 41 = 581]$. Therefore they would represent this number as YYYYYYYYY <<<< Y (i.e. the symbol for nine in the 60s column, and the symbol for forty-one in the units column).

In a similar way, for the Babylonians, 3730 was not three thousands, seven hundreds, three tens and 0 units, but 1 three-thousand-six-hundred, 2 sixties, and 10 units $[(1 \times 3600) + (2 \times 60) + (10 \times 1) = 3600 + 120 + 10 = 3730]$. Therefore, a Babylonian would write: Y YY < .

---

15. If you write out all the numbers between one and a hundred inclusive, how many ones will you have written?

It took a while for other civilisations to catch up with the Babylonians, but they got there in the end. The Incas arrived in Central and South America in the 13th century and established a huge empire with a complicated administration. To keep track of all their wealth, and who owed them tax, they used knotted lengths of cord, called *quipus*. A *quipu* would have a series of knots tied on to it. The first group of knots would record units, the second group would record tens, the third group hundreds, and so on. Each *quipu* would contain numerical information about some aspect of a town's life, and they were collected and kept by an official, called the 'Rememberer'.

Other systems have been created in other places over the centuries. The Chinese came up with one, the Mayans have decorated their temples with their symbols for numbers, and alternatives have been created in modern times as well. Computers and other machines use the binary place-value system to operate. In this system, the values of the 'columns' are based on powers of two. Each column has a value that is two times greater than the previous one. So the first column (reading from right to left) is for units, the second for 'twos', the third for 'fours', the fourth for 'eights', and so on.

Using this system, a computer considers the number that we write as 46 to be made up of 1 thirty-two, 0 sixteens, 1 eight, 1 four, 1 two and 0 ones:

$$(1 \times 32) + (0 \times 16) + (1 \times 8) + (1 \times 4) + (1 \times 2) + (0 \times 1)$$
$$= 32 + 0 + 8 + 4 + 2 + 0$$
$$= 46$$

So, in binary, this number becomes 1 0 1 1 1 0. And 129 is thought of as 1 one-hundred-and-twenty-eight, 0 sixty-fours, 0 thirty-twos, 0 sixteens, 0 eights, 0 fours, 0 twos and 1 one [$(1 \times 128) + (1 \times 1) = 128 + 1 = 129$]. Therefore, in binary, our 129 becomes 1 0 0 0 0 0 0 1.

As you can see, there were plenty of options for how to write down numbers, and plenty of different civilisations who tried them out. But eventually, in Europe at least, the merchants admitted defeat, put away their abaci, and started scribbling sums on bits of paper. It was the beginning of written arithmetic.

8 BORROWING AND CARRYING

Imagine the trauma for the average shopkeeper of 14th century Europe: his neighbour and rival has switched from doing sums on the abacus to this new-fangled paper-and-pencil method, and is serving his customers twice as fast. He has got to keep up, or his business will fold. But he doesn't understand all these symbols and lines, or the need for 'borrowing' or 'carrying', and there are dark rumours about the dangerous powers of these new numbers. Some of his friends say that they are a plot by Eastern sorcerers to bewitch the minds of honest Western folk. They say that a mysterious king, called Algorismus, has laid down that sums must be calculated according to his particular 'laws', but that this is really just a trick to ensnare his mind. What to do?

There is no arguing with economic reality. So he takes himself off to one of the new schools, where they teach how to use these strange numbers. His fellow students laugh at him when he tells them about the fearsome Algorismus. They inform him that his friends are in fact talking about al-Khwarizmi, the famous Arab mathematician, and that many of the new techniques he is learning come from his book. They say that any standard system of doing a particular type of sum is called an 'algorithm' in English, in memory of this great mathematician. He breathes a sigh of relief, starts scribbling sums in his brand new blue exercise book, and dreams of the day that his teacher gives him a gold star for getting all his long multiplications right.

> 16. 'Twice four and twenty blackbirds
>   Were sitting in the rain.
>   Jill shot and killed a seventh part.
>   How many did remain?'   [Wells (1992) page 126]

You will be pleased to know that not much has changed in the world of addition (or subtraction) on paper over the years. This is what English mathematics scholar Robert Recorde (1510–1558) had to say on the matter in his *The Ground of Artes*, written in 1543. It is the first ever mathematics textbook, and you will probably recognise the feeble attempt to dress the sum up as a realistic problem:

'What? Addition is very easy to do, me thinketh I can do it even now. There came through Cheapside two droves of cattle, in the first was 848 sheep, and the second was 186 beasts… [to calculate the total number of sheep] I must write one number over the other… so that the first figure of one be under the first figure of the other…when you have so done, draw under them a straight line…now begin at the first places towards the right hand always, and put together the two first figures of those two sums, and look what come of them, write under them, right under the line…'

The imaginary reader in the *Ground of Artes* is quite happy with this explanation and the similar one for subtraction, only mentioning his confusion when he has to deal with carrying and borrowing. These two processes rely on the base ten number system that the Arabs came up with. If the 14th century shopkeeper tried to work out how many sheep were wandering through Cheapside, and wrote the sum down as instructed, he would find this on the paper in front of him (without the headings of the columns):

	H	T	U
	8	4	8
+	1	8	6

Carefully following the instructions of his 14th century teacher he adds the two figures in the last column and tries to write down the number fourteen beneath them. The teacher happens to be looking over his shoulder at this point, and goes ballistic. He barks at him to write down the four and 'carry' the one, whilst standing uncomfortably close to him.

There is a reason for this strange request. By adding the last two numbers, the shopkeeper has in fact been counting the number of units in the two numbers combined. In total there are fourteen units, but this is equivalent to one ten and four units. Therefore he writes down a four in the units column for the total, and a little one in the tens column to remind him to include it when he counts up how many tens there are in the two numbers combined.

He bends over his book studiously, forehead wrinkled earnestly, and works out that there are thirteen tens in total in the two numbers combined (including the one that he has 'carried'). In a moment of inspiration, he realises that this is equivalent to one hundred and three tens, and writes down a three in the tens column, and a little one in the 'hundreds' column.

Finally, he calculates that there are ten hundreds in total (including the one 'carried'), quickly reasons that this is equivalent to one thousand and zero hundreds, and writes down a zero in the hundreds column and a little one in the thousands column. He looks around, fails to see any more thousands, and writes a big one in the thousands column. His answer is 1034.

---

17. As a result of discovering a batch of two-headed fish, and five-legged chickens, the citizens of a town panic, and there is a mass evacuation. The population of the town falls from 3652 to 2759. How many people fled?

The concept of 'borrowing' works in reverse. Say that, of the first 848 sheep wandering through Cheapside, 186 of them stopped off to do some Christmas shopping, and that you, as the shepherd, wanted to know how many were left in your flock. If you followed the instructions in *The Ground of Artes*, you would write down the following sum:

	H	T	U
	8	4	8
−	1	8	6

Just as in addition, you start the calculation from the right hand side. You work out that eight sheep arrived in Cheapside, but that six have stayed on to shop, and so of these sheep, only two are left for you to look after. You then turn your attention to the tens column, and discover that forty sheep arrived in town, but that eighty have wandered off. This is a problem, until you realise that there are eight hundred more sheep that started the day with you. If you 'borrow' one of these hundreds, you can say that 140 sheep arrived in town, of which eighty have left, leaving sixty that remain. Finally, you turn to the hundreds column. You have 'borrowed' one of the hundreds, so there are only seven hundred sheep left to think about, of which one hundred are now purchasing last-minute gifts. So there are six hundred further sheep for you to care for. In total, you have 662 sheep left in your flock.

9 LONG, LONG MULTIPLICATION

Long multiplication is a different kettle of fish altogether. Most civilisations have avoided it as being too much trouble. Computers don't go anywhere near it, preferring to work out all sums by high-speed addition (or subtraction). However, those that have taken it on have come up with all sorts of ways of dealing with it.

The technique that we use is a fairly recent arrival on the scene, and became popular because it is easy to print. It relies on 'splitting up' a multiplication sum into more manageable calculations. So, 2641×275 is done in this way:

$$
\begin{array}{r}
2\ 6\ 4\ 1 \\
\times\ 2\ 7\ 5 \\
\hline
1\ 3^{3}2^{2}0\ 5 \\
1\ 8^{4}4^{2}8\ 7\ 0 \\
+\ 5^{1}2\ 8\ 2\ 0\ 0 \\
\hline
7^{1}2^{1}6^{1}2\ 7\ 5
\end{array}
$$

In the above example, the sum 275×2641 is considered to be the same as $(5 \times 2641) + (70 \times 2641) + (200 \times 2641)$, since '275 lots of 2641' is the same as '5 lots of 2641, plus 70 lots of 2641, plus 200 lots of 2641'. The first row of the addition (13 205) is the result of 5×2641, the second row (184 870) is the result of 70×2641, and the third row (528 200) is the result of 200×2641. You find the final answer (726 275) by adding together the answers of each of these multiplications.

This method simplifies matters because, in each of the three multiplications, it is only necessary to multiply 2641 by a single-digit number, which is a fairly pain-free process. 5 × 2641 gives the first row.

70 × 2641 gives the second row, but this can be considered as 10 × (7 × 2641). 7 × 2641 is a single-digit multiplication (Answer: 18 487), and multiplying by 10 just means you have to add a zero at the end of your answer for 7 × 2641 (because the digits have each moved one column to the left).

200 × 2641 gives the third row, but this can be considered as 100 × (2 × 2641). 2 × 2641 is 5282, and multiplying by 100 just means that you have to add two zeros at the end of this answer (because the digits have each moved two columns to the left).

All long multiplications using this technique follow the same pattern, which is why your teacher was always shouting at you to add a zero at the end of the second row, two zeros at the end of the third row, three zeros at the end of the fourth row, and so on, and so on, and so on.

---

18. In the following sum, each letter represents a different number from 1 to 6 inclusive.

$$
\begin{array}{r}
A\,B \\
\times \quad C \\
\hline
D\,E\,F
\end{array}
$$

Can you work out what the sum must be?

The most popular method for working out what we call 'long multiplications' in ancient times was called 'gelosia', after the grills which were placed over the windows of houses where nuns or chaste women lived. There is nothing more upsetting than an attractive chaste woman being stuck on the other side of a grilled window, but that has nothing to do with the mathematical process, which only went out of favour because it is so cumbersome to print.

Take the problem 258 × 24. This is a three digit number multiplied by a two digit number, so you draw a 3 by 2 rectangular grid, and write the three digit number above it, and the two digit number to the side:

The next step is to draw in some diagonal lines and continue them a little, like this:

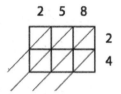

You then simply fill in each box by multiplying the two associated numbers, putting the 'tens' above the diagonal, and the 'units' below, and add down the diagonals created by the lines, 'carrying' in the normal way, as shown below:

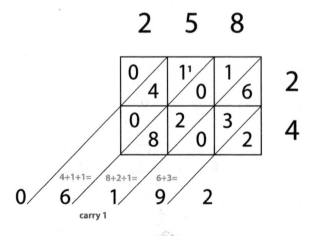

In this problem, when adding the numbers in the 'third' diagonal, the answer was 11, and so '1' was written beneath the diagonal, and '1' carried to the next diagonal. Reading along the bottom of the grid from left to right, you get 6192, which is the answer. This method is a little tricky to start with, but once you've got the hang of it can be used to multiply any large numbers.

For example, take the sum 346 × 229. This multiplication will require a 3 × 3 grid, but, in all other respects, the technique is the same as the last problem:

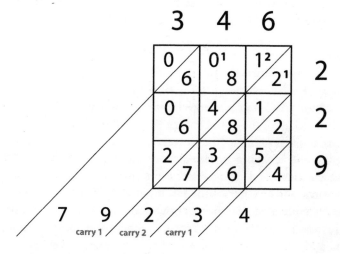

Answer: 79234.

19. Some of the numbers have gone missing in this gelosia problem. Work out what they must have been.

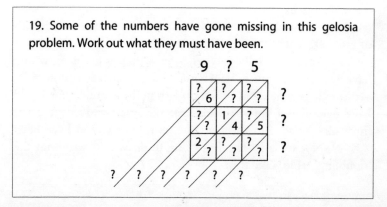

There are more than two ways to skin a cat, though. The ancient Egyptians solved any multiplication sum using a 'doubling' technique. Whatever the number they were multiplying, they would write out a 'doubling' table of it, and then use it to solve the problem.

The Rhind Mathematical Papyrus was discovered in the middle of the 19th century in the ruins of a small building close to the temple of Rameses II in Thebes. It caught the eye of Alexander Henry Rhind, who happened to be spending the winter in Egypt due to health reasons, and was fond of browsing through the second-hand shops in Luxor in search of cheap presents for his relatives.

The papyrus turned out to be the work of a scribe called Ahmose. He had written it some time around the middle of the 16th century BC, but was copying from a work written during the second half of the 19th century BC. Ahmose had high hopes for his book, calling it 'the correct method of reckoning, for grasping the meaning of things and knowing everything that is, obscurities…and all secrets.' I thought this sounded quite promising, and so I felt a little let down when I discovered that it mostly contained examples of how to divide different amounts of bread and beer between different numbers of men. But it does contain the earliest preserved riddle: 'There are seven houses each containing seven cats. Each cat kills seven mice and each mice would have eaten seven ears of spelt. Each ear of spelt would have produced seven hekats of grain. How many hekats of grain are lost?'★

An Egyptian schoolboy, when faced with this problem, would have been able to concentrate better than his modern counterpart, because there would have been no girls in the room. So he would have figured out fairly quickly that the cats would have eaten forty-nine mice. But the next multiplication would have been 7×49. Not so easy. The first thing to do was to write out the 'doubling' table. Like this:

★ Wells (1992) p3.

1	49	1 × 49
2	98	2 × 49
4	196	4 × 49
8	392	8 × 49
16	784	16 × 49
32	1568	32 × 49

The next stage was to combine different 'rows' to get the answer to the required sum. Here, he wanted 'seven lots of 49'. 'Seven lots of 49' is the same as 'one lot of 49, plus two lots of 49, plus four lots of 49'. And, according to the doubling table, 'one lot of 49' is 49, 'two lots of 49' are 98, and 'four lots of 49' are 196. Therefore, the Egyptian schoolboy added 49 and 90 and 196 to get the answer to the multiplication: 335.

In fact, he could use the same idea to get any multiple of forty-nine. 'Three lots of forty-nine' is the same as 'two lots and one lot'. 'Twelve lots of forty-nine' is the same as 'four lots and eight lots'. 'Forty-three lots of forty-nine' is the same as 'thirty-two lots and eight lots and two lots and one lot'. Any multiple of forty-nine can be found by adding together a combination of the rows of a doubling table like the one above.

And I am not just talking about multiples of forty-nine. The same Egyptian schoolboy would have gone about the problem of 258 × 24 in exactly the same way. Here is his 'doubling' table:

1	258	1 × 258
2	516	2 × 258
4	1032	4 × 258
8	2064	8 × 258
16	4128	16 × 258

In this case, the schoolboy wants '24 lots of 258', which is the same as '8 lots of 258' and '16 lots of 258'. From the table, '8 lots of 258' is 2064, and '16 lots of 258' is 4128. So, the answer to the problem is 2064 + 4128, which is 6192.

Bernadette makes a final note in her exercise book, underlining it with her gold highlighter, and carefully packs away her things into the various different compartments of her executive briefcase. She waits behind her desk, as Mr Barton dismisses the class row by row, and files out dutifully when her name is called. There is something she wants to look up in one of the reference books in the library.

Charlie watches his classmates leave the room enviously. Outside, patterns of children are already forming, scattering and regrouping in the various games taking place on the tarmac and worn patches of grass. Girls cluster in huddles to weave their conspiracies. Teachers roam about drinking cups of tea, talking about whatever it is that teachers talk about, and turning a blind eye to the ritual humiliation of a small boy in the corner of the playground who has the misfortune to have been christened Richard Head.

Charlie sighs and concentrates on doing the homework he failed to hand in. He feels weighed down by the injustice of his life. It is not his fault that his younger brother thought that a page of long multiplication sums would be a suitable final resting–place for his pet goldfish, which had died, like many before it, as a result of receiving a massive overdose of fishfood. Charlie had tried to persuade his brother to exhume the fish, so that he could recover the homework, but to no avail. The suggestion had only resulted in an outburst of pain and grief.

Mr Barton sits at his desk, picking at his teeth with a long, pale forefinger. He is unconcerned at Charlie's loss of liberty. He has never really understood why people like to keep pets, or why they tolerate the disorder and chaos that animals bring into their lives – and he certainly has no sympathy for a boy who uses his homework to wrap up a dead fish. He shuts the door to drown out the racket from the playground outside, and busies himself with marking the perfect columns of figures in Bernadette's pristine exercise book.

10 LONG DIVISION EXPLAINED

So finally to the nightmare that is long division. Who would have thought that one sum, neatly written at the top of the page, could give birth to such sprawling columns of numbers? However, before trying to make sense of these numbers, it is worth looking at a couple of other techniques, because they hold the key to why long division is, in fact, a perfectly sensible thing to do.

First of all, it is necessary to realise that every division sum can be considered as a subtraction. $63 \div 9$ can be transformed into the question: 'How many nines make up sixty-three?' and there is no better way of figuring this out than working out how many times you can take nine away from sixty-three until you have nothing left.

$$63 - 9 = 54$$
$$54 - 9 = 45$$
$$45 - 9 = 36$$
$$36 - 9 = 27$$
$$27 - 9 = 18$$
$$18 - 9 = 9$$
$$9 - 9 = 0$$

In other words, $63 - 9 - 9 - 9 - 9 - 9 - 9 - 9 = 0$. You have to take away seven lots of nine to reduce sixty-three to nothing, and so the answer to the initial division sum is seven. Any division sum can be solved in the same way. $24 \div 6$ is equivalent to finding how many sixes make up twenty-four. $24 - 6 - 6 - 6 - 6 = 0$. You have to remove four sixes to reduce twenty-four to zero. The answer to $24 \div 6$ is 4.

This might seem a long-winded way of going about division, but it becomes much more useful when it is speeded up a bit and applied to more difficult problems. Take the problem $336 \div 24$. You can solve this in the same way as the problems above by working out how many times you have to take away twenty-four from three-hundred-and-thirty-six before nothing is left. $336 - 24 - 24 - 24 - 24 - 24 - 24 - 24 - 24 - 24 - 24 - 24 - 24 - 24 - 24 = 0$. The answer is fourteen.

---

20. What is the least number into which each of the first nine numbers (1–9) will divide exactly?

However, it would be good to get to this answer a bit more quickly, so instead of subtracting one 'twenty-four' at a time, you start subtracting several 'twenty-fours' in one go. In order to do this, you just have to know some multiples of twenty-four that are easy to work out. In this case, we are going to use the facts that '10 lots of 24' are two-hundred-and-forty, and '2 lots of 24' are forty-eight.

Back to the problem of $336 \div 24$. With this new information, you don't have to subtract each 'twenty-four' individually. You can first subtract 240, which you know is equivalent to '10 lots of 24':

$$
\begin{array}{r}
336 \\
-\ 240 \quad (10 \times 24) \\
\hline
96 \\
\hline
\end{array}
$$

Once you have subtracted '10 lots of 24', 336 has been reduced to 96. Next, you can subtract 48, which you know is equivalent to '2 lots of 24':

$$\begin{array}{r} 96 \\ -\ 48 \\ \hline 48 \end{array} \quad (2 \times 24)$$

After this subtraction, you have reduced the initial number, 336, to 48. You can then subtract another 48 (which is still equivalent to '2 lots of 24'), and the 336 is all gone:

$$\begin{array}{r} 48 \\ -\ 48 \\ \hline 0 \end{array} \quad (2 \times 24)$$

Overall, in order to reduce 336 to 0, you have taken away '10 lots of 24', followed by '2 lots of 24', followed by a further '2 lots of 24'. In total, you have taken away '14 lots of 24', and therefore the answer to the problem, 336 ÷ 24, is 14. More importantly, you arrived at this answer in only three steps.

The same technique can be applied to all divisions. 336 ÷ 12 is equivalent to the question: 'How many twelves must you take away from 336 before nothing is left?' To answer this quickly, by using the same methods as before, you need some multiples of twelve that are easy to calculate. You can choose whichever ones you like, and then start subtracting. One way to solve the problem is this:

$$\begin{array}{r} 336 \\ -\ 120 \\ \hline 216 \\ -\ 120 \\ \hline 96 \\ -\ 60 \\ \hline 36 \\ -\ 36 \\ \hline 0 \end{array} \quad \begin{array}{l} \\ (10\ \text{lots of }12) \\ \\ (10\ \text{lots of }12) \\ \\ (5\ \text{lots of }12) \\ \\ (3\ \text{lots of }12) \\ \\ \end{array}$$

In total, you have subtracted '28 lots of 12', and so the answer to the original division is 28.

However, there is no fixed route to the answer. You could also solve the same problem in the following way:

$$
\begin{array}{rl}
336 & \\
-\quad 240 & \quad \text{(20 lots of 12)} \\
\hline
96 & \\
-\quad\ \ 96 & \quad \text{(8 lots of 12)} \\
\hline
00 &
\end{array}
$$

It is all a matter of personal taste.

---

21. You are part of a very organized protest against the reintroduction of bears into the Scottish Highlands. There are 252 protesters, and you form eighteen columns of equal length in order to make your point. How many people are there in each column?

Now that you know that it is possible (and quite sensible, in fact) to think about division as repeated subtraction, you should be able to deal with the tangled web of numbers that make up a 'long division', because any long division is at heart a formalised and simplified form of repeated subtraction.

In general, in the mathematics classroom, division is nearly always connected to sweets (or possibly marbles), so let's say that the problem is to divide 3968 sweets fairly between sixteen people. You take a deep breath and set out the division problem as shown below:

$$16\overline{)3968}$$

The first step is to see how many times sixteen divides into thirty-nine, and write down the answer (2) above the 9. After this, you need to work out what 16×2 actually is and write this down below the 39. The figures are already beginning to crawl, but the sum should now look like this:

$$
\begin{array}{r}
2 \\
16\overline{)3968} \\
32
\end{array}
$$

Remember that this division sum is equivalent to the question: 'How many sixteens must I take away to reduce 3968 to zero?' At this point, you have identified the fact that it is possible to take away '200 lots of 16' from 3968, and that '200 lots of 16' is 3200. The zeros have been left out in order to simplify the layout of the sum.

The next step is to subtract the 32 from the 39, and 'bring down' the 6, like so:

$$
\begin{array}{r}
24 \\
16\overline{)3968} \\
-\,32 \\
\hline
76
\end{array}
$$

Here, you are really working out how much of 3968 is left, after you have subtracted '200 lots of 16'. The actual answer is 768 (3968 − 3200), but for the purposes of the division, it is not necessary to write the whole of this answer down. All that is required are the first two digits.

The process of long division is then repeated on 76. 16 goes into 76 four times. The four is written above the 'box', and the actual result of 4 × 16, is written beneath the 76:

$$
\begin{array}{r}
24 \\
16\overline{)3968} \\
-\,32 \\
\hline
76 \\
64
\end{array}
$$

Again, you have identified that from the 768 sweets remaining, it is possible to subtract '40 lots of 16', and that '40 lots of 16' is 640. Once more, the zeros have been left out.

Next, you subtract 64 from 76, write down the answer beneath, and 'bring down' the eight:

$$
\begin{array}{r}
24 \\
16\overline{)3968} \\
-32\ \\
\hline
76 \\
-64\ \\
\hline
128 \\
\end{array}
$$

You have in fact worked out that after a further '40 lots of 16' (or 640) sweets have been subtracted from the original number, you are left with 128.

$$
\begin{array}{r}
248 \\
16\overline{)3968} \\
-32\ \\
\hline
76 \\
-64\ \\
\hline
128 \\
-128\ \\
\hline
0 \\
\end{array}
$$

In the final stage of the division, you identify that it is possible to subtract a further '8 lots of 16' from this remaining total, and that once you have done this, you have reduced 3968 to zero. In the course of the long division, you have subtracted '200 lots of 16', then '40 lots of 16' and lastly '8 lots of 16'. In total, you have subtracted '248 lots of 16', and therefore the answer to the problem is that everyone can have 248 sweets. The whole problem could be expressed in the same way as the earlier problems that you solved using repeated subtraction, and it would look like this:

```
      3968
   − 3200        (200 lots of 16)
   ───────
      768
   −  640        (40 lots of 16)
   ───────
      128
   −  128        (8 lots of 16)
   ───────
      000
```

You can see that there is not a whole lot of difference between the two ways of dealing with the problem.

---

22. At a particular petrol station, for every four tokens you collect, you receive a fluffy dice and a token in return. If you have collected forty-eight tokens, how many fluffy dice can you claim?

Just to hammer the point home, here is another division. The solution by each of the above methods is given side by side:

```
        3562
   14 |49868          49868
   −   42           − 42000        (3000 lots of 14)
   ──────           ───────
        78             7868
   −    70           − 7000        (500 lots of 14)
   ──────           ───────
        86              868
   −    84           −  840        (60 lots of 14)
   ──────           ───────
        28               28
   −    28           −   28        (2 lots of 14)
   ──────           ───────
         0                0
```

Life is never easy, so it is likely that, on the few occasions in your life it is necessary to share out a bag of sweets fairly between several pairs of grubby hands, you will discover that there are a few sweets left over. For example, if you try and share out 3970 sweets between sixteen children, you will find that each child can have 248 sweets, but that there are two left over at the end.

In the emotionally sterile world of division, this 'two' is called the remainder. But this knowledge will not quiet the angry group of hungry creatures in front of you. So, in this scenario, what does the remainder mean? It means that, in order to complete the division, you are going to have to get involved with fractions. The easiest way of calming the rabble is to take each sweet and divide it into sixteen exactly equal pieces. Each piece will be 1/16 of a sweet, and each child will receive two pieces (one from each of the remaining two sweets). Therefore, if you play by the rules, each child should receive 248 whole sweets and a further 2/16 of a sweet, but it is probably easier just to eat those two sweets yourself.

---

23. The inhabitants of a small village in the North of England (population 125) had nothing better to do than to create the world's largest sausage, which measured 4018cm in length when completed. On dividing it up, the villagers gave 50cm of it to the village elder, and the rest was split equally amongst themselves. How much sausage did each villager receive?

---

11 CHECKING IT ALL ADDS UP

It is human to make mistakes, especially when you are struggling through something like a long division. So it is natural to want to know if the end product of your work is correct. Of course, in the classroom, there was no need to reassure yourself. Your teacher marched threateningly among the desks, with his red pen ready to dispense justice. He flashed the pen through the air inches from your ear, and with a few flicks of his wrist, he passed judgement on hours of calculation. A nonchalant tick for correct work, and a savage slash through mistakes.

However, there are ways of checking calculations yourself to see if they are correct, without relying on the hidden knowledge of a red pen. The most common method is called 'casting away nines'.

Say you wanted to check that your addition of 848 and 186 was correct. For each number, you add the digits, and take away nine as many times as is possible. So, for 848, you add 8 and 4 and 8 to get 20, and then start taking away nines: $20 - 9$ is 11, and $11 - 9$ is 2. This is in fact exactly the same as finding the digit sum, where you add all the digits of the number, and then add the digits of the answer that you get, repeating the process until you get a single-digit number. In this case, $8 + 4 + 8$ is 20, and $2 + 0$ is 2.

Similarly, for 186. By the method of casting away nines, $8 + 6 + 1$ is 15, then take away nine to get 6. Or alternatively, by the digit sum method, $1 + 8 + 6$ is 15, and $1 + 5$ is 6. Next, you need to go through the same process with the answer to the sum (which is 1034) to get its digit sum as 8. Since the calculation was an addition, you add the digit sums of the numbers added, and you should find that this result is the same as the digit sum of the answer to the calculation. In this case, the digit sum of the first number is 2, and the digit sum of the second is 6. Add 6 to 2 to get 8, which is the same as the digit sum of the answer.

You can use the same method for all other calculations, except division. If you want to check a subtraction, you subtract the digit sum of the second number from the digit sum of the first number, and check to see if the result is the same as the digit sum of the answer you have got, whilst, for multiplications, you multiply the digit sums.

Just to prove that I am not lying, 217 × 43 is 9322. The digit sum of 217 is 1, the digit sum of 43 is 7, and the digit sum of 9322 is 7. 7 × 1 is 7, and so the check suggests that I have calculated the sum correctly.

However, before you get too excited, I ought to mention that this method is not entirely foolproof. There is a small possibility that, by sheer misfortune, you will make a mistake in a calculation that leads to an answer that happens to have the same digit sum as the correct answer. All I can say is that nothing is perfect, and that if you check a calculation in this way, and the check works, you have very probably got it right.

If this state of affairs is just too imprecise for your needs, then you can always check a calculation by reversing it. For example, you have carefully calculated that 154 + 289 is 443, and you want to make sure that you are correct. Therefore, you just as carefully work out the sum 443 − 289. If you find that the answer to this subtraction is 154, then your initial calculation was correct − it was. In the same way, in order to check that 1589 − 297 is 1292, you work out the sum 1292 + 297. If the answer to this addition is 1589, then all is well with the world.

Multiplications and divisions can be checked in this way as well. If you think that 123 × 56 = 6890 (it doesn't), the most certain way of checking your answer is to divide 6890 by 56. In this case, the initial calculation was wrong, and so 6890 divided by 56 does not give the expected answer of 123. You need to check your working for the initial problem. Similarly, if you think that 11913 ÷ 57 = 209 (it does), the best thing to do is to multiply 209 × 57. You ought to get 11913, which proves that your calculations are mistake-free.

Of course, it is perfectly possible to avoid doing long written calculations nowadays, and there is no shame in it. The lady behind the checkout desk does not write down the price of each item as it sails past her, and then carefully add up the neat columns of figures in front of her. There is no flurry of long multiplication in the bureau de change, when you change pounds into euros. No – there are machines to take care of things, smooth over the cracks, and leave the memory of troubled schooldays undisturbed.

But it isn't a bad idea to have a rough idea of how much the items on the supermarket conveyor belt should come to, or how much the black-market money-exchanger should give you, or what the figure should be at the end of your bank statement. At least, that is my excuse for sitting in a small back room in my house, happily adding up all the cheques I have paid into my account, and subtracting all the withdrawals. After all, the bank could have made a mistake – although I have to admit that I have never yet found one. Anyway, the hours spent in that back room have nothing at all to do with a bizarre satisfaction in going though a massive calculation to discover that my figures match with the little black digits at the bottom of the last page.

Still, the sanctuary of a back room is different to the hurly-burly of a curry house on a Friday night. There is not the necessary peace or required time to spend in a long analysis of how many poppadoms you have consumed. Speed is of the essence. And so, in general, a quick rounding of the items on the bill will be enough to check that a mistake hasn't been made. For example, if you had eaten six poppadoms at 90p each and two curries (which cost £4.85 and £6.25), a quick check would involve estimating the cost of one poppadom to be £1, and the cost of the two curries to be £5 and £6 respectively. You would expect a bill of around (6 × £1) + £5 + £6 = £17.

Alternatively, you could just add up all the pounds, and leave the pence alone. Or, if you wanted to be even more accurate, you could add the pounds and then add on fifty pence for each item consumed (as a likely average of pence for all the items). So, for the curry-house example above, you look at the bill and see that you owe ten whole pounds for the two curries, and then, instead of actually calculating how many pence you owe, you estimate 50p for each item. Therefore, you figure that you owe roughly £10 + (8 × 50p), which is £14. If the bill is much more (or less) than this, you ask to see the manager. No need to be more complicated than that. Certainly no need to get out a pen and start doodling on the back of your paper napkin.

> 24. I enter a small village shop in rural Rwanda to buy four litres of palm oil. The owner reaches behind the counter to pull out a big vat of oil, but apologises and says that she has only an old three-litre paint-tin or a five-litre lubricating-oil container in which to pour out the measures. You tell her not to worry, because it is possible to measure out four litres using these two objects, even though they are not marked in any other way. How do you do it? (You are allowed to pour liquid back into the vat of oil.)

That is a way of checking addition, and the same strategies work to check subtractions. But when changing pounds into euros (or vice versa) you need to check the man's multiplication, especially if this is some kind of shady better-than-the-official-exchange-rate kind of situation.

Let's say you are in the dingy corner of a Dover pub, where a man with an eye-patch is offering you 1.43 euros to the pound, and you want to change 19 pounds. A sensible check for you to do, would be to multiply 1.50 by 20, which is 30. Therefore you want about 30 euros, although you expect a little less because you rounded up both of the numbers in the calculation (i.e. you don't quite have 20 pounds, and the exchange rate wasn't quite 1.50), and so you have slightly over-estimated how much you should get.

Similarly, if you wanted to change 22 pounds at a rate of 1.62 euros, you would quickly multiply 20 by 1.5, and figure out that you should get a little more than thirty euros, because both numbers were rounded down (i.e. you have more than 20 pounds, and the exchange rate is higher than 1.5).

If you round up one figure, and round down the other figure, it is not easy to tell whether you have overestimated or underestimated. You will just have to trust the man.

That is one difficult financial situation out of the way. But imagine you were coming back from your holiday, and you found yourself in the decrepit booth of a run-down Calais restaurant facing the same man, with 32 euros in your pocket. The rate is 1.43 euros to the pound. This time you will have to subtly divide 30 by 1.5, and expect around 20 pounds.

Overestimate or underestimate? The actual calculation that you would have done on your calculator (if you hadn't left it in your small back room with your bank statements) would have been 32 divided by 1.43. Your approximate calculation was 30 divided by 1.5. You rounded down the 32, so there was less money to be divided, which you would expect to lead to an underestimate. And you rounded up the 1.43, so that less money had to be divided between a larger number, which leads to an even bigger underestimate. The man in the booth should give you more than 20 pounds.

Similarly, if you were changing 37 pounds at a rate of 2.21 euros to the pound, you would round up the 37, round down the 2.21; quickly figure out 40 divided by 2, and hold out an expectant hand for around 20 pounds. But this would be an overestimate, because your approximate sum divided more money than you actually had by a lower exchange rate.

Again, when dividing, if you round up both numbers or round down both numbers, you can't expect to know whether you have made an underestimate or an overestimate. And if all this dividing makes you feel unsteady on your feet, you could always ask the man to tell you how many pounds you get for one euro, and go back to multiplying. Really, he ought to have told you in the first place.

So there you go. That's it. The world of written calculations has been safely negotiated. No need to avoid long multiplication and long division any more. Like all good tricks, there is a rational explanation of them. If only they had told you that from the start...

PART TWO

DIFFERENT KINDS OF NUMBER

1   KIT-KATS AND KOSHER

It is time to look into the shadows of the mathematics classroom, where fractions, decimals and percentages lie covered in dust. These three topics are closely connected to one another, because they are all different scales that we use for counting. For example, if you are counting sheep, ½ a sheep is equivalent to 0.5 of a sheep or 50% of a sheep, and 1 ½ sheep are equivalent to 1.5 sheep or 150% of a sheep. Since they are all types of scale, it is best to look at how scales in general work, before you deal with the individual topics. And this brings us into the territory of ratio and proportion…

At school, it never became clear what the difference between ratio and proportion was. In fact, it was quite hard to work out what either of them actually were, even if you happened to stumble across one of them on their own. They seemed to apply vaguely to some situation where you might be called upon to share some sweets unfairly between several people, so that one of them got a massive pile of candy, and everyone else was left with a miserable handful. Or for a particular moment in a supermarket in some far-flung suburb, when you were aware of the exact number of apples, and of the fact that there were twice as many oranges as apples, but you couldn't quite recall how many oranges there were.

Generally speaking, ratios are used when we are describing the relationship between different parts of the same general whole. So we talk of mixing different colours of paint in the ratio 3:2 to create a paint of a new tone, or we speak of sharing out a pile of sweets between two people in the ratio 2:1. In both these cases, paints and sweets can be thought of as parts of a whole. The paints are combined to make a new paint. The shared sweets are parts of the initial pile.

Proportions are used to describe the connection between two distinct things. So we can say that the amount of tax to be paid is proportional to the amount of money you earn. Or you can claim that sexual prowess is proportional to the length of your nose. In both these cases, the things compared are quite different from one another. They are not part of a more generalised whole.

The problem with those explanations is that there are plenty of cases where the boundaries between proportion and ratio are blurred, and the two words become interchangeable. It is possible to talk about sharing sweets in a ratio of 2:1, but it is also possible to say that the amount of sweets that I have is proportional to the amount of sweets that you have. In this case, the word you use depends on the way you think about the sweets. In the first case, you are focused on the fact that the sharing process is dividing up a group of sweets into two separate parts. Therefore, the two groups of sweets that result are closely connected in your mind to the one original pile of sweets – they are parts of a whole. In the second case, you are simply focused on comparing the number of sweets in each of the piles that lie in front of you. The two piles are separate and distinct, and there is no thought of a past in which they were united as one large happy pile. As a result, in the first case you talk of ratios, and in the second case you talk about proportions. It is all in your head.

Still, whatever the precise distinction, both ratio and simple proportion deal with situations where there is a numerical connection between two different groups of things. They deal with the type of situation where you know that if you double one thing, then you double the other. Like currency exchange. If ten dollars are worth six pounds, then you can be pretty sure that twenty dollars are worth twelve pounds. Or dogs and legs. If four dogs have sixteen legs, then eight dogs have thirty-two legs (assuming that all dogs have the same number of legs).

> 25. I have spent many hours trying to perfect the preparation of beans in Rwanda. I have finally worked out that to provide properly for two people, it is necessary to boil 200g of beans with four onions, six tomatoes, three tablespoons of brown sugar, two tablespoons of mixed spice and three teaspooons of salt. On Christmas day, I had to cook enough beans for eleven people. What quantities of ingredients did I need to use?

Proportion and ratio pop up all over the place. In music, it is vital that string lengths are of the right proportion. If you take two strings of the same tension, one of which is twice as long as the other, and pluck them, they will produce two notes an octave apart. Other notes are produced by strings with lengths in a whole-number ratio with the original string. A string one-and-a-half times the length of the original one, when played with it, will produce a perfect fifth. Two strings with lengths in the ratio 4:3 will play a perfect fourth. And so on. Of course, it is only possible to check the truth of this, if you have any idea of the difference between a perfect fourth and a perfect fifth. I don't.

Ratios can also tell you what to eat. According to Jewish religious law certain foods, such as pork, are forbidden. If you are making dinner, and accidentally mix in a piece of pork, then as long as the ratio of pork to kosher food is 1:60 or less, the dinner can still be eaten. If not, the food must be thrown out. But this rule only applies if you drop the pork into your cooking accidentally. I am not quite sure what the method for finding the exact ratio of pork to food is. I suspect that once it has been carried out, the food may have lost much of its appeal, and the issue of whether it is kosher or not may have become irrelevant. I wouldn't be so keen on a meat stew if my father had been rooting around in it for half-an-hour.

Proportions also have a large influence on our collective ideas about beauty. You might not think that rectangles are particularly beautiful things, but draw one. Just a quick doodle. Don't think about it too much. If you measure the sides of the rectangle that you have drawn, I am willing to wager that the length will be roughly 1.618 times as long as the width. This ratio (1:1.618) is called the Golden Ratio (or the 'Divine Proportion', or the Golden Mean, or the Golden Section). It crops up all over the place. It can be found in the seed and petal arrangements in trees and plants, and composers often use it to divide up pieces of music. In mathematics, 1.618 is called Phi (φ), allegedly after the ancient Greek architect Phidias, who regularly used the Golden Ratio in his buildings. It appears to be the connection between lengths of things that we find the most attractive. Not too fat, and not too thin.

The ancient Greeks used it in building the Parthenon. The front view of the temple has its height and width in the Golden Ratio, although whether this was intentional on the part of the Greeks, or more of an intuitive decision, is open to debate.

Besides being built to the aesthetically pleasing proportions of the Golden Ratio, the Parthenon has other clever tricks included in its design. For example, the floor of the temple curves very slightly upwards as it leads away from the entrance to make the interior seem bigger than it really is, and the columns holding up the roof become thicker higher up to stop them from looking thinner higher up, if you know what I mean. These tricks were all intended to make the building appear more symmetrical. The Greeks were pretty obsessed with symmetry.

> 26. You have a cat and a dog. One day you weighed them, and found that the dog weighed 20kg and the cat weighed 5kg. They visited you in a dream and told you that they wanted you to share out the pet food in the same ratio as their weights. How much food should each get from a tin that contains 575g?

English philosopher Francis Bacon said, 'There is no excellent beauty that hath not some strangeness in the proportion.'★ But scientists tend to disagree. Faces that conform to our general ideas of attractiveness will fit neatly inside a rectangle with lengths that conform to the golden ratio. Plastic surgeons probably use

---

★ Bacon (1985) p189.

this technique in constructing beautiful people. Who knows what happens when the anaesthetist has knocked you out? They probably put away all the impressive looking gadgets, and whip out a rectangle made from cardboard and double-sided sticky tape.

Oh yes, and most credit cards have proportions which are very close to the golden ratio. And so do Kit-Kats (the four-fingered chocolate variety).

So, given that it is responsible for so much, it is unsurprising that proportion has been viewed in the past with religious awe. Pythagoras and the Pythagoreans believed that numbers were the very essence of the universe, and that all relationships could be expressed numerically. It was Pythagoras who first discovered that music was made by strings whose lengths were in whole-number ratios. His discovery was based on his observation that when a blacksmith struck an anvil, the note produced differed with the weight of the hammer used.

He also observed that there were five planets (they hadn't discovered the other ones yet), and that they appeared to move along orbits which seemed to be connected by similar ratios to musical notes. There was only one conclusion: simple whole numbers, and the connections between them, were the fabric of the cosmos. They were the instructions to explain the universe we live in, the pattern which decides our place in it, and the music to which it dances.

Sadly, the whole theory came crashing down, when it was discovered that the connection between the diagonal of a square and its side is not expressible as a proportion in whole numbers. Or, to explain it in another way, if you divide the length of a diagonal of a square by the length of its side, you do not get a number that can be written as a fraction. You get $\sqrt{2}$, which is one of those never ending decimals that do not repeat themselves. Shame.

---

27. You are traveling to work by train. You notice that the ratio of people who look happy to be alive to people who have forgotten how to smile is 2:7. If there are thirty-six happy people, how many people should consider looking for a better job?

2 A 'RYCHE SHEPEMASTER'

Proportion hasn't gone away since the days of the Pythagoreans. It is very much part of everyday life. You use it to work out how much currency you should expect when you change money. You use it to change from kilometres to miles. You use it to work out the price of your brussel sprouts at 78p per 100g. You use it to change a recipe for four people into a recipe for six.

I most recently encountered the perils of proportion in my daily life when I entered my local curry house with my new girlfriend and some friends, and sat down to my standard Friday-night meal. As always I avoided eating poppadoms at the beginning, in order to be able to fit in a peshwari naan later on. There were no surprises. The evening went along as merrily as the many similar evenings that had gone before it on other Friday nights.

Until the bill came. I did not check the calculation on the back of a paper napkin, as was usual, because I did not want my girlfriend to know about this dark side of my personality. (The checking would take place later, at home, in the sanctity of my study.) But I did realise that this bill was not as simple as the ones from my bachelor past. Because, gentleman that I am, now I was paying for two. Suddenly, the bill required more than a simple division. I had to work out how much one person should pay, and then double it. And that is proportional thinking. It is just one more proof of the fact that relationships make life more complicated.

28. Here are six tumblers, three full and three empty, arranged in a row:

What is the smallest number of moves needed to leave the tumblers alternately full and empty? Every time a tumbler is picked up, that counts as a move.

So given that proportion is waiting behind the most unexpected corners, how should you deal with it when it turns up? We have already seen that people have been tackling proportional problems for a long time, and so it made sense for different cultures to come up with a mechanical way of solving such problems. The methods that they found are all largely similar, and so they are all generally referred to as the 'Rule of Three'.

The Rule of Three was known in China as early as the 1st century AD. It appears in ancient Indian texts of the 5th century AD, in the works of the Islamic mathematicians from the 8th century AD onwards, and in the first European textbooks. People were generally very enthusiastic about it because it solved so many useful problems. The Europeans of the Renaissance called it the 'Golden Rule', and the Indian mathematician Bhaskara II, wrote that it pervaded the world of mathematics just as the god Vishnu pervaded the universe through his countless manifestations.

Robert Recorde, in *The Ground of Artes*, introduces it as follows: 'I…wyll teache you…the rule of proportions, whiche for his excellencie is called the Golden rule?' He then goes on to demonstrate it, in true mathematics-teacher fashion, by means of several examples in the hope that his reader will grasp the general principle. The first of the examples is as follows:

'If you paye for your borde for three monthes sixteen shillings, how much shall you paye for eight monthes?'

He tells the reader to lay out the information given in the following way:

The numbers on the left-hand side must be the same type of thing (for this example, months), and each of these numbers is paired in the same row with the piece of information connected to it. In this case, you know that three months' lodging costs sixteen shillings, and you want to know how much eight months' lodging costs (so the space for this information is left blank).

With the numbers ordered in this way, Recorde then says: 'And now to know my question, this must I do: I muste multiplye the lowermost on the left side, by that on the right side, and the sum that amounteth, I must divide by the highest on the left side'. In other words you start with the bottom-left number (8) and, following the lines of the figure, multiply by the top-right number (16). When you have worked this out, you divide your answer by the top-left number to get the solution to the initial problem. For any problem of this type, as long as you set out the information in the same way, you can follow this method unthinkingly and arrive at the correct answer.

---

29. A train takes four seconds to enter a tunnel which is 2km long. If it is travelling at 160 km/h, how long will it take to pass completely through the tunnel?

But it is best not to put such blind faith in the abilities of your teacher. How do you know such methods are not some government-sponsored plot to keep you in the dark about how much rent you should be paying? So let us try to unravel the logic behind Recorde's numbers and lines.

The trick with this problem is to find out how much you pay for one month's board. Once you have calculated this, you can quickly find out how much the lodging will be for any number of months by multiplying. The process can be represented well in a table:

Months	Shillings
3	16
1	$16 \div 3 = 5\ 1/3$
8	$8 \times 5\ 1/3 = 42\ 2/3$

Given the initial information that three months' board costs sixteen shillings, it is possible to calculate how much one month's lodging costs by dividing sixteen by three. Once you have this piece of information you can calculate the cost of any length of stay by multiplying by the number of months. In this particular problem, you are asked to find out how much eight months' board will cost, so you multiply by eight. You can see that the calculation that you carry out is the same as the calculation that Recorde's instructions for the 'Rule of Three' asked you to do, except you have done the division ($16 \div 3$), before the multiplication [($16 \div 3) \times 8$]. The resulting answers are the same.

Recorde quickly moves on to look at more complicated problems, but they can all be tackled using the same kind of thinking. Here is another one:

'There is supposed a lawe made that (for furthering of tillage) every man that doth keep shepe, shall for every ten shepe eare and sowe one acre of grounde, and for his allowance in sheepe pasture there is appointed for every four shepe one acre of pasture. Nowe is there a ryche shepemaster whyche hath 7000 akers of grounde, and would gladlye kepe as manye sheepe as he might by that statute. I demaunde howe many shepe shall he kepe?'

There are no printing errors in the passage above. It is just that people were not so fussy about spelling back in Recorde's day. You just spelt the word sheep according to how you personally felt that it should be spelt at that particular moment.

There are plenty of different approaches, but here is one way of solving the problem, which relies on the same methods as were used previously. Here comes the table:

Sheep	Acres to eare and sow	Acres of pasture	Total
10	1		
4		1	
20	2	5	7

The problem states that for every ten sheep (or shepe), a farmer should have one acre of land to 'eare and sowe', and for every four sheep, he should have one acre of pasture. Twenty sheep is the lowest multiple of four and ten. Using the previous information, if a farmer has twenty sheep, he should have 2 × 1 acres of land to 'eare and sowe', and 5 × 1 acres of land for pasture. This means that for twenty sheep, he should have a total of seven acres of land. But this 'ryche shepemaster' has 7000 acres of land to play with, so:

Sheep	Acres to eare and sowe	Acres of pastures	Total acres
20	2	5	7
20 000			7000

Since 7000 acres is one thousand times more land than seven acres, the farmer is able to look after 1000 × 20 sheep. That is a lot of sheep. He will certainly have his hands full, especially since sheep in those days were a lot more aggressive than they are now. As the 'scholar' in Recorde's book points out: 'the shepe are waxen so fierce nowe and so myghtye, that none can withstande them but the lyon.'

30. On the last Saturday of every month, all Rwandan citizens must work for the good of the country (a system called Operation Umuganda). If forty Rwandan citizens can plant four hundred saplings in five hours, how many saplings can forty Rwandan citizens plant in six hours?

3 PROPORTION HAS ITS PROBLEMS 1

Of course, you have to be careful. You can't just go around applying this kind of proportional thinking willy-nilly. There are plenty of situations where it is not appropriate. There are plenty of times when it is not true that if one thing doubles, so does another. If this is the case, you cannot use the types of approach that we have been talking about. For example, it is often not true that if you buy twice as much of something, you pay twice the amount, because of discount deals for bulk purchases.

It is especially common to assume that length and area, and length and volume, are proportional. In other words, people often think that, if you double the lengths of a shape, you double its area, and that if you double the lengths of an object, you double its volume.

However, if you have ever made a mistake of this sort, you can console yourself with the knowledge that you have proved the eternal existence of the soul. At least Socrates, the famous ancient Greek philosopher, uses a mistake of this sort by a slave boy as part of his argument to prove this concept in a dialogue by Plato, called the 'Meno'. He draws a picture of a two by two square in the dust, and asks the boy its length and area.

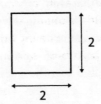

The boy correctly answers that the length of a side is two and its area is four. Then Socrates asks what the area of a square twice as big will be, and the boy responds, again correctly, that it will have an area of eight. Socrates' next question is what the length of the side of the larger square will be. And the boy replies: 'Clearly, Socrates, it will be double.'

The boy has assumed that length is directly proportional to area. He has assumed that doubling the area of the shape involves doubling the lengths. Socrates goes on to demonstrate that a square with a side of length four units has an area of sixteen. The boy becomes very confused, and admits he no longer has a clue about what the length of the side of a square of eight units should be. I am sure that you recall a similar feeling of bewilderment when the words coming out of your teacher's mouth no longer had any meaning or connection with the world that your mind inhabited.

Socrates, however, is pleased with his work, and observes that it is better for the boy to know that he does not know the answer to the question, than to think that he knows what is in fact the wrong answer. He goes on to draw another picture in the dust:

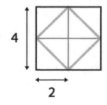

By asking a series of questions about the diagram, Socrates helps the boy to mend the error of his ways. The boy is happy to agree that the area of the blue square is sixteen square units. He has already agreed previously that this is the area of a square with a side of length four units. By examining the picture he reaches the conclusion that the red square contains half the area of the blue square, and so must have an area of eight square units. After a bit of further thought, he realises that the sides of the red square are in fact the diagonals of 2 × 2 squares. Therefore, he concludes that the length of a side of a square with an area of 8 square units is the same as the length of a diagonal in a 2 × 2 square. He is happy to leave it at that – he has work to do – but he could have gone on to use Pythagoras' famous theorem to calculate the exact length of a diagonal of a 2 × 2 square. It is around 2.8 units.

As a result of this episode, Socrates concludes that, since he did not actually give the boy any information, the information must have been in the 'soul' of the boy in the first place, and that since the boy has never been taught the information, his soul must have picked it up somewhere else, before the boy came into being. From these facts, Socrates deduces that the soul must be immortal, and that it must contain all knowledge. The problem is that this knowledge fades over time, and must be reawakened periodically, just as Socrates has reawakened the slave boy's faded knowledge of geometry. It is a shame that not all our mathematical mistakes can lead to the foundation of significant philosophical theories.

---

31. In a dawn raid, a group of savages carry off your beautiful new wife. When you finally wake up – you are notoriously deep sleeper – they have already travelled forty miles. You set out in pursuit, and chase them for a distance of twenty-three miles, but give up in the belief that you are not able to catch up with them. When you turned back, you had cut the lead of the savages to thirty-two miles. If you had carried on, how many more miles would you have travelled before you caught up with them?

4 PROPORTION HAS ITS PROBLEMS 2

Over the years, plenty of classic mathematical problems have involved proportion. One particular type of problem involved calculating how long it will take for a basin to fill up, given that the various pipes that feed it provide water at different rates. The first instances of problems of this type appear in ancient Chinese mathematical works, but from then on, they are found in the works of mathematicians of various different cultures. The following is the earliest known example – it is from the 'Chiu Chang Suan Shu', the most influential of the ancient Chinese mathematical works, which was written some time after 300 BC by an unknown author.

'There is a basin filled by five channels. The first channel alone fills the basin in 1/3 of a day, the second channel alone fills it in a day, the third channel alone fills it in 2½ days, the fourth channel alone fills it in 3 days, and the fifth channel alone fills it in 5 days. If water comes through all the channels at once, how long does it take for the basin to fill?'

There are various different ways of solving this problem, but here is a method that makes use of the kind of proportional thinking that we have relied on previously. Again, here is a table containing the information you have been given:

	Channel 1	Ch 2	Ch 3	Ch 4	Ch 5
1/3 day	1 basin				
1 day		1 basin			
2 1/2 days			1 basin		
3 days				1 basin	
5 days					1 basin

The trick is to find a common multiple of the different numbers of days that each channel takes to fill the basin, and then work out how many times each channel would fill the basin in this time. In this case 15 is a common multiple of 1/3, 1, 2 ½, 3 and 5, so you need to look at how many times each channel would fill the basin in this time.

	Channel 1	Ch 2	Ch 3	Ch 4	Ch 5
1/3 day	1 basin				
1 day		1 basin			
2 1/2 days			1 basin		
3 days				1 basin	
5 days					1 basin
15 days	45 basins	15 basins	6 basins	5 basins	3 basins

The results for fifteen days are arrived at by using simple proportions. For example, Channel 1 fills the basin in 1/3 of a day. Therefore, it would fill the basin three times in one day, and forty-five times in fifteen days. Channel 3 fills the basin once in 2 ½ days, and so it would fill it twice in five days, and six times in fifteen days. The other results are reached by similar calculations.

Therefore, if all the channels work together to fill the basin, then in fifteen days they would fill the basin 45 + 15 + 6 + 5 + 3 = 74 times over. Once you have discovered this, you can use proportional reasoning again to find out how many days it will take for all the channels to fill it just once:

15 days	74 basins
15 ÷ 74	1 basin

It takes all the channels 15/74 of a day (which works out as approximately 292 minutes) to fill the basin once.

Solving proportional problems is not just a deviant way of passing the time; it can earn you the respect and admiration of important people. Archimedes was an ancient Greek mathematician. He was excellent at DIY, and his special interest was designing war machines to beat the Romans, who kept on attacking his home town of Syracuse. He came up with massive catapults to hurl rocks at their ships, giant pincers to grab them and topple them over, and powerful mirrors that concentrated the rays of the sun to set them alight. Or so the stories say.

However, back then, as now, it was important to get noticed. You had to grab your chance. Archimedes was just a common-or-garden craftsman plying his trade at the court of King Hiero, putting up shelves and grouting ceilings, when his chance came. The king had commissioned a statue, on top of which was to be a golden crown. He gave the gold for the crown to an artist, but when the artist came back with the crown he had made, the king suspected that he had actually mixed some silver in with part of the gold, and kept the rest of the gold for himself. However, the king was unable to prove that this was the case.

Enter Archimedes. He had only just had time to pull on a pair of Y-fronts (and not his best pair at that), after running naked around the streets of Syracuse in celebration of his discovery of the fact that the volume of water displaced by an object is the same as the volume of the object itself. Now he was ready to put the discovery to good use. The exact details of the problem are not known, but let us say that he measured the weight of the crown to be 2 kg, and found that it had a volume of 20 cm^3 (by putting it in a tub of water, and measuring how much water spilled out). He then took 2 kg of gold and 2 kg of silver, and discovered that they displaced a volume of 15 cm^3 and 30 cm^3 respectively.

From this information, Archimedes was able to calculate exactly how much gold the craftsman had used to make the crown, and exactly how much gold that he had kept for himself to use on a boozy weekend in Ayia Napa. The king was very impressed, Archimedes' fame and fortune was assured, and finally he could put some clothes back on.

I thought you might want to see if you could match Archimedes for intellect, and so I have left out the details of his possible solution of the problem here. However, it is not easy trying to contend with a genius, and so here are a couple of hints to start you in the right direction.

Assuming that the crown is made up of a mixture of gold and silver, then let the weight of the gold be w_1 and the weight of silver be w_2. From the information it is possible to write down two equations. The first comes from the fact that you know that the crown weighs 2 kg. The second comes from the fact that you know the crown has a volume of 20 cm^3.

It is harder to work out what the second equation is. You need to work out the volume of a weight of gold, w_1, and the volume of a weight of silver, w_2. The table below might help you to find the volume of the weight of gold, w_1:

Weight of gold (kg)	Volume of gold (cm^3)
2	15
4	2 × 15 = 30
1	1/2 × 15 = 17 1/2
2/3	1/3 × 15 = 5
w_1	?

Once you have found the two equations, then I am afraid that you are going to have to solve them as simultaneous equations. Sorry.★

★ See Part Three, Chapter 5 for more on solving simultaneous equations.

I think we will leave it there for the moment. Some people at the back of the class are getting restless. Charlie is concentrating on focusing the sun's rays via his watch face onto Mr Barton's retina, and simultaneously throwing chewed pellets of paper at the back of Bernadette's neck. The bell is about to go anyway.

But Bernadette has other ideas. Ignoring the sharp needles of pain caused by Charlie's missiles, she raises her hand straight up in the air, and wiggles her fingers to attract Mr Barton's attention. You have to admire her courage (or is it just insensitivity to the feelings of her classmates?).

Bernadette has been doing some extra mathematics in her spare time, and she has come across a problem that she is unable to do. This is more than she can bear, and she won't rest until she has found a satisfactory explanation of how to do it. She hands it over to Mr Barton, who takes it, inspects it, and writes it up on the board in his almost illegible hand-writing. 'If six mowers doe mowe forty-five acres in five daies, howe many mowers will mowe three hundred acres in six daies?'

Mr Barton makes some unpleasant noises in the back of his throat, snorts loudly several times, and then starts scribbling a solution. This problem is more complicated than previous ones, because it contains an element of inverse proportion. If six mowers, mow forty-five acres in five days (to use modern spelling – although you don't see many human mowers nowadays), then three mowers will take ten days, and twelve mowers will take two-and-a-half days. That is – if you double the number of mowers, you half the amount of time taken, and if you third the number of mowers, you triple the amount of time taken.

Mr Barton had only just had time to explain this point to a largely unresponsive class, when the bell went. Chairs scraped, desks slammed, books shut, and the rest of his explanation was lost in the general confusion. Only Bernadette, who had, of course, remained in her seat, caught the rest of it. But I bumped into her in the library recently, and she explained to me how to solve the problem.

There are three variables in this problem: the number of mowers, the number of acres mowed, and the number of days taken to mow them. To start off with, it is best just to focus on two of them, and leave the other unchanged. So let us just worry about the number of mowers and the number of days it takes them to mow forty-five acres:

Number of mowers	Number of acres mowed	Number of days taken to mow
6	45	5
1	45	30
5	45	6

We start off with the initial information given in the problem: six mowers will mow forty-five acres in five days.

The problem is requires that you find out the number of mowers needed to mow a certain area in six days, so it might be worthwhile finding out how many mowers it takes to mow forty-five acres in this period of time. As we have already noted, if you increase the number of mowers, then you decrease the number of days taken to mow the same area.

It will be useful to find out how long it will take one mower to mow forty-five acres, because from this fact, it will be possible to calculate how long it will take any number of mowers. If six mowers take five days, then it will take one mower six times as long (i.e. thirty days). But if one mower takes thirty days, then, in order to mow the same amount of land in six days (i.e. a fifth of the time), you will need five times as many mowers. It will take five mowers six days to mow forty-five acres.

The problem requires that you find out how many mowers are needed to mow three hundred acres in six days, and so now you must play around with the number of acres to be mowed. But the difficult part is over – you have dealt with inverse proportion.

The rest of the solution relies on direct proportion. The more mowers you have, the more land they can mow. So:

Number of mowers	Number of acres mowed	Number of days taken to mow
5	45	6
5/45 = 1/9	1	6
300 × 1/9 = 33 1/3	300	6

Once again, it is a good first step to find out how many mowers it takes to mow one acre in six days. Since, in this period of time, five mowers mow forty-five acres, then, in the same period of time, 1/9 mower will mow one acre (don't worry about the unsettling concept of 1/9 mower for now). Therefore, to mow three hundred acres in six days, you will need 33 1/3 mowers.

33 1/3 mowers is a strange number. You can't have a third of a human mower. It is either all or nothing. So perhaps you should just employ thirty-four mowers, and explain that every day one of them can go home after doing only 1/3 of his hours. Or employ thirty-four mowers, one of whom you know is incredibly lazy, and works three times as slowly as all the others. I don't know – it is up to you.

5   COLOURING IN PIZZAS

In the classroom of Mr Barton, thirty heads are bent over thirty orange exercise books next to thirty chunky textbooks on top of thirty graffiti-covered desks. There is an example on the board, and it shows clearly the method used for dividing fractions. The majority of the children in the class are painstakingly applying this method to the columns of problems in the textbook. Bernadette turns over to a new page in her exercise book with a self-satisfied sigh. She has already finished Exercise A, and is looking forward to the harder challenges expected in Exercise B.

Mr Barton looks out over his kingdom, and smiles inwardly at the tranquil scene in front of him. He looks fondly through his thick lenses at the neat lines of calculations in Bernadette's book, stopping to place a few more ticks next to her work. He patrols the room with a firm and measured step, noting with satisfaction the furrowed brows of the children as they struggle to multiply together large numbers using only pencil-and-paper methods. Mr Barton does not allow calculators in his class. They spoil the fun.

But over in the back corner, all is not well. Charlie has stopped writing in his book, and is leaning backwards in his chair. Mr Barton moves swiftly over and bends close to its dirty pages, which are partially obscured by jam-stains and the bodies of crushed insects. With obvious distaste, he begins to read Charlie's latest offerings, but soon stops with a look of total incomprehension and reaches over to place a series of angry red crosses beside his work. 'Look at the board, Charlie,' he mutters. 'Look at the board! You don't just divide the bottoms and divide the top. No, no, no…you flip the second one upside down, and then multiply tops and bottoms. It's easy. Think boy think!' He walks away muttering angrily.

Charlie shakes his head in confusion. All the rules keep getting confused in his mind. He knows that to add and subtract fractions, there is some rule about finding the lowest number that the bottom number in each fraction goes into, and that multiplying is easy because you just multiply the bottoms and the tops. And now, he has been told that there is a whole different set of rules for dividing fractions. But how is he to remember which is for what? And why are they all different anyway? It just doesn't make sense. He begins to correct his work. What was it he had been told to remember…something about turning inside out and 'quantifying'?

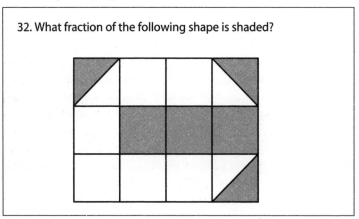

32. What fraction of the following shape is shaded?

Charlie is not the first person to shake his head in confusion. Fractions have always been a problem. They seem such a sensible idea to start with. I can remember thinking that 'fractions' was just another name for 'colouring in'. It was soothing – carefully scribbling away with my dark blue crayon until I had shaded in three squares out of four. I tried desperately hard not to go over the lines, but there was always one moment of carelessness when my crayon spiralled out of control and left a messy blue loop dangling off the edge of my colouring.

Most people have no problem with understanding the concept of a half, or a quarter, or in fact any single fraction. They are essentially a useful set of scales for when you start cutting up things. You share a pizza between five people, so you cut it into five equal pieces. Each piece is one-fifth of the pizza, and you can count the number of pieces using connected fractions.

If someone is on a diet, and someone else is greedy, the greedy person can take two pieces of pizza and end up swallowing two-fifths of it. If the greedy person is also rude, he might lunge across the table and grab a third piece from one of his friends. Then he will eat three-fifths of the pizza. If the greedy person decides further to alienate his friends, he might steal a fourth piece off the plate, and then he will have eaten four-fifths of the pizza, and if he is in fact a socially maladjusted psychopath, he might scoff the whole pizza. In this case, he will have eaten five-fifths of the pizza, which is the equivalent of eating one whole pizza.

Once you have created this thing called a 'fifth', you can extend the scale as far as you like. Six-fifths is a whole pizza and another one-fifth piece. Twenty-fifths is equivalent to four whole pizzas. And so on. You don't have to stick to pizzas. The scale can be applied to anything. You just need to remember to divide it into five EQUAL pieces. A fifth of a metre. A fifth of a second. A fifth of a book. A fifth of a monkey…

There are an infinite number of other scales to describe pieces of things. You create them by cutting things up into different numbers of equal pieces. For example: you have reached the limit on your credit card for the umpteenth time, and, wanting to escape the depressing cycle of debt and interest, you cut the card into twenty equal pieces. Not only have you solved your financial problems (possibly), but you have created twentieths out of your credit card (definitely). If you come to regret your decision, and carefully start to stick the pieces back together again, only to give up after you have stuck together seven pieces because you realise that the card will never work again, then you have in your hands seven-twentieths of a credit card.

Things can get a bit more complicated if you want them to. Instead of cutting up a credit card, or a pizza, you might want to cut up two pizzas. You can cut up two pizzas into quarters (i.e. divide two pizzas into four equal parts), and you will find that a quarter of two pizzas turns out to be half of one pizza. Or you can cut up four pizzas into eighths (i.e. divide four pizzas into eight equal parts), and discover that an eighth of four pizzas is also a half of one pizza. When you try to calculate with fractions it is these kinds of complications that can begin to leave you stumped.

> 33. Is it possible to write down a number that is equivalent to 100 using just four 9s (you cannot use operation signs like + or −)?

6 WHAT THE EGYPTIANS DID

If you are looking to lay the blame for fractions at anyone's door, then place it in a small package containing an explosive device at the door marked 'Division'. People in Babylon, China and Egypt had all happily mastered how to add, subtract and multiply numbers in whatever form they had developed them. Fresh from this success, I am sure that they turned to the process of dividing them without any worries at all. After all, division was only the opposite of multiplication. What could go wrong? So imagine their horror when they discovered that some divisions 'don't go'. Imagine their annoyance when they found that division meant that you have to come up with a completely new kind of number. It must have been as if they had sensibly put aside an hour or two to buy a special kind of halogen light bulb for their kitchen, only to discover that it was out of stock at their local hardware store, and they had to trek all the way to the superstore way out of town. They had done everything right. In the high-tempo world of an ancient civilisation, it was best not to cram too many errands into one day (what with the lack of wheels and lower life expectancy), and division was a lengthy business. But this discovery of fractions was unforeseen, and now all they had to look forward to were traffic jams and an argument with the trolley superintendent.

You can sense exactly this frustration in the way the Egyptians reacted to the issue. They refused to accept any fractions other than fractions with a numerator (the top bit) of one (we call such fractions 'unit fractions') – with the exceptions of 2/3 and ¾.

This testy restriction to unit fractions ended up giving the Egyptians a massive headache. It meant that if they wanted to divide seventeen pies between ten people, they could happily give everyone one pie, but splitting the remaining seven fairly became a real problem. They could not give everyone 7/10, because 7/10 was not a number they understood. Nor for some reason could they write down that each person got a 1/10 + 1/10 + 1/10 + 1/10 + 1/10 + 1/10 + 1/10, because in their annoyance at ending up at the out-of-town superstore they decided to ban the writing of the same fraction several times in this way. In fact, they came up with very particular rules about using unit fractions. To make a fraction like 7/10, they had to add together the least possible number of different unit fractions, and they always wrote them down by starting off with the largest. So, for them, 7/10 was actually 1/5 + 1/2.

There were two direct results of this strange method of dividing up pies. Firstly, specialised scribes toiled away to create division tables for problems like the one above, so that everyone else could live in peace. And secondly, at a birthday party, the hosts always made sure they had the same number of cakes as there were guests, and it was very bad form to fail to turn up without warning. The resultant trauma of working out how to divide up the cakes could last for days...

---

34. This weekend, I shot a quarter of the squirrels in my garden, whilst you shot half the squirrels in your garden, but the Animal Rights Activists put a petrol-bomb through my door, because I was the bigger murderer. How is that possible?

---

7   EQUIVALENT FRACTIONS

As we have seen, when it comes to fractions, you are actually dealing with a whole set of different scales. You use different ones depending on the particular pieces that the world throws at you. When watching soccer matches, you think in terms of halves, but when watching basketball, you think in terms of quarters. If you are employed by the Narcotics Squad, you should be familiar with eighths and sixteenths, and if it is your job to count the number of dwarves looking after Snow White, it is best to be familiar with sevenths.

Everything is fine as long as you stick with one scale at a time. But as soon as you come across problems that involve using more than one scale, all hell breaks loose. This is exactly what happens when you start trying to add fractions, or subtract, or multiply, or divide them.

Before you read on, just remember that the Egyptians were very clever people and they had all sorts of problems with what follows. They were not alone among earlier civilisations. The Romans also struggled to get their heads around the concept, only managing to come up with a limited number of words for the different fractions that appeared in everyday business transactions. Their basic unit of weight was called an *as*. Every as contained twelve *uncia*, and so their words for fractions tended to be connected with twelfths. 1/12 was called a *deunx*, 6/12 was called a *semis*, and 1/144 (a twelfth of a twelfth) was called a *scripulum* (which sounds more like a painful medical condition than the name of a number). As a result of the fact that they represented fractions only in word form, they found it very hard to do any calculations with them at all.

In fact, it was not until around 500 AD that anyone came up with a way of handling fractions in a form that we would recognise. The Indians were the first to get there, thanks to their invention of the decimal place-value number system. The Indian method for writing fractions was transmitted to Europe via the Arabic world, where the habit of separating the two numbers in a fraction with a line was introduced. This line had a special name. It was called a vinculum – a very friendly-sounding word for an object that has filled so many students with dismay over the years. The Indian fractions first appeared in the Western world in the works of the Italian mathematician, Leonardo of Pisa, (also known as Fibonacci) in the early 13th century, but their use was not widespread until much later than this.

---

35. How can you measure a length of 15cm if you only have a pencil, an unmarked ruler of 7cm and an unmarked ruler of 11cm?

So, given all the problems and difficulties that people faced in coming up with the concept of a fraction, you deserve some credit for even trying to tackle them. First up: equivalent fractions. These are groups of fractions that look different but are in fact the same. To find a fraction equivalent to the one you have already got, 'you just have to multiply the numerator (top bit) and the denominator (bottom bit) by the same number'. So, if you start off with 2/3, and multiply the numerator and the denominator by 2, you get 4/6, which is in fact the same as 2/3. If you multiply top and bottom by 3, you get 6/9, which is also the same. If you multiply top and bottom by 4... I think the penny should have dropped by now.

At school, I probably spent at least half an hour writing out all the fractions equivalent to 2/3, but as soon as I got home, the technique of 'timesing top and bottom by the same number' vanished from my head. This is because, if you just deal with the numbers, it's not obvious why it's true. To show why it is obviously true, you need pictures. And the best kind of picture for this purpose is a picture of a chocolate bar. At least, your teacher called it a chocolate bar, but it nearly always turned out to be exactly the same as a rectangle.

If you draw a rectangle, divide it into thirds, and shade in 2/3, you might draw something like this:

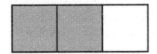

Now, if you split each column in the rectangle into two equal parts, you double the number of columns and you double the number of shaded columns. You have shown that 2/3 is exactly the same as 4/6, and at the same time made the rectangle look a little bit more like a chocolate bar:

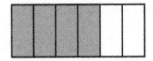

If you split each column in the original rectangle into three equal parts, you triple the number of columns, and you triple the number of shaded columns. You have shown that 2/3 is the same as 6/9:

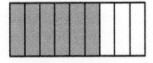

And if you split each column into four equal parts, you show that 2/3 is the same as 8/12:

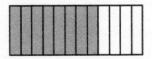

In fact, you could split each column into as many equal parts as you liked, and show that 2/3 is the same as an infinite number of other fractions, but you should probably stop now.

It is possible to draw a rectangle to represent any fraction you want, and to work out which fractions are equivalent to it in the same way. The process of splitting the columns of the original chocolate bar into equal pieces is equivalent to 'timesing the top and the bottom by the same number'. The rules do make sense after all.

---

36. In a survey, forty-one out of sixty French people said that they believed that the English were a bunch of hooligan thugs who thought that the highest form of cuisine was sausages with tomato ketchup. In another survey, thirty-three out of fifty Englishmen said that they believed that the French were selfish snobs who butchered frogs and snails for fun, and didn't put up enough of a fight in the war. Is there greater love for France in England, or greater love for England in France?

---

8   ADDING FRACTIONS ON PAPER

If I remember rightly this is how I was taught to add (and subtract) fractions when I was at school:

1) You find a number that both the denominators of the fractions go into.
2) For each fraction, you multiply the numerator and the denominator by whatever number you must multiply the denominator by to get the number in (1), and replace the original fractions by the fractions that result.
3) You add (or subtract) the numerators of these new numbers, and leave the denominators unchanged to get the answer to the sum.

For example, to add 2/5 and 1/3:
1) You find a number that both 5 and 3 go into (such as 15).
2) For 2/5, you must multiply 5 by 3 to get 15, and so you multiply the top and the bottom of the fraction by 3 to get the equivalent fraction 6/15.

   For 1/3, you must multiply 3 by 5 to get 15, and so you multiply the top and the bottom of the fraction by 5 to get the equivalent fraction 5/15.
3) You add the top numbers of these two new fractions (5 + 6 = 11). The answer to the sum is 11/15.

The silent question is 'Why?' Why is this how you add fractions together? Say you want to add two fractions like 2/5 and a ¼.

You can represent those two fractions by folding pieces of paper and shading in the relevant sections. In this situation one piece of paper is equivalent to one 'whole'.

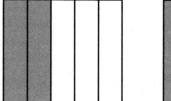 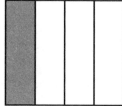

The problem is that you want to know how much of one piece of paper the combined shaded areas on the separate pieces of paper represent. But you have no way of comparing fifths with quarters. You do not know how many fifths make up a quarter.

You can solve this problem by taking the first piece of paper and folding it horizontally into quarters and taking the second piece of paper and folding it into fifths. Like this:

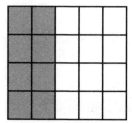 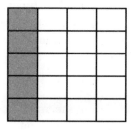

By folding the paper into fifths and quarters, you have folded it into twentieths. Twenty is a number that both five and four go into (in fact it is the lowest possible number – also known as the Lowest Common Multiple). This explains part one of the rule. In general, when adding any two fractions using this folding technique, after you have folded the pieces of paper for the second time, the number of sections that you get will be a multiple of the bottom numbers of the fractions that you are adding together.

You have split the two fifths into four equal parts, so that the two fifths are the same as eight twentieths. Likewise, you have split the quarter into five equal parts, so that the quarter is the same as five twentieths. This is the same as multiplying both numbers in '2/5' by four, and both numbers in '1/4' by five. That is where the second part of the rule comes from, since four and five are the numbers which multiply the denominator of each fraction respectively to give twenty.

Finally, the problem asks you how much of one piece of paper would be shaded in if you combined 2/5 and ¼. From the diagram you can see that it is 13/20. You have added the top numbers from '8/20' and '5/20', and that was what the third part of the rule told you to do.

---

37. Three candidates were running for Office. One of them got half the vote. Another got two fifths of the vote. What fraction did the third candidate get?

So there you go. You can add (or subtract) any fractions you like using this technique, and each stage in it is equivalent to one of the steps in the rule I was given at school. For example, if you want to subtract 1/6 from ¾, you fold two pieces of paper to represent these two fractions:

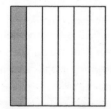 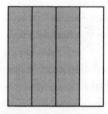

You then fold the first paper horizontally into quarters, and the second paper horizontally into sixths, which divides both pieces of paper into twenty-fourths. In this case, twenty-four is NOT the Lowest Common Multiple of 4 and 6, but it is certainly a number that they both go into. Now it is clear that 1/6 is equivalent to 4/24 and that ¾ is equivalent to 18/24. You have multiplied the top and bottom of each fraction by four and six

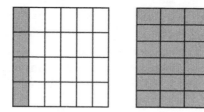

respectively. To find the solution to the problem, you subtract 4/24 from 18/24 to get 14/24, which is equivalent to 7/12. For both the paper-folding method and the method I was taught at school, it is fine, at the first step, to use a number that is not the Lowest Common Multiple of the denominators of the two fractions you are working with. It just means that, at the end, it will be possible to find a simpler fraction that is equivalent to your solution. In this case, you used 24 (a multiple of 4 and 6) rather than 12 (the Lowest Common Multiple of 4 and 6), to get the answer 14/24. It is then necessary (mainly for the sake of appearances) to simplify your final answer to 7/12, although both fractions represent the same amount of rectangle. (If you had used the paper-and-pen method, and picked 12 as the number that both 4 and 6 go into at the first stage, you would have got 7/12 as your answer straight away.)

> 38. A lion is incarcerated in a pit which is twenty metres deep. Every day the lion manages to climb ½ metre up the side of the pit, but every night it slips back down by 1/3 metre. How many days will it take for the lion to escape the pit?

I am not suggesting that you should use the paper-folding technique if you ever have to add some fractions in the course of normal life (which is very unlikely). The paper-and-pen method is quicker. But it is nice to know that 'the rule' makes sense. It is not some random sequence of steps. It is not the mental equivalent of being told to 'Docey-doe' and then 'Right Hand Star' at a barn dance by a man with a beard and a microphone.

9 TURN IT UPSIDE-DOWN AND MULTIPLY

Why stop at adding? What about multiplying fractions? Given the complicated rule for adding fractions, it is a bit of a surprise to discover that to multiply them 'you just times the top numbers and times the bottom numbers'. In fact, it is more than surprising, it is downright suspicious.

Once again, it is easiest to think about this sort of problem in terms of rectangles. 'What is $1/5 \times ¼$' is the same question as 'what is $1/5$ of a $¼$?' (This follows from the nature of multiplication. '2×3' is equivalent to '2 lots of 3', and '3×4' is equivalent to '3 lots of 4'. In the same way '$1/5 \times ¼$' is equivalent to '$1/5$ lots of $¼$', or just '$1/5$ of a $¼$'.) You can represent a $¼$ on a rectangle like this:

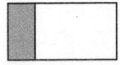

But you want a 1/5 of a 1/4, which would look like this:

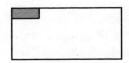

Next, you want to know what fraction of the whole rectangle this is. To know this you have to divide all the other quarters into fifths as well, so that you have divided the shape into equal sections. Like this:

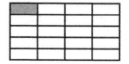

Now you have 4 × 5 = 20 equal little rectangles. The total number of little rectangles will always be the same as the number you get when you multiply together the denominators of the two fractions. In this particular problem you want one rectangle in one of the quarters, and so the answer is 1/20.

But if the problem had been 1/5 × ¾ (or 1/5 of ¾), you could solve it in exactly the same way, except you would want one rectangle from each of three of the quarters, and the solution would be 3/20.

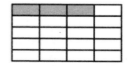

Finally, if the problem had been 3/5 × ¾ (or 3/5 of ¾), you could go back to the same shape, but this time you would want three rectangles from each of three of the quarters, and so you would end up with 3 × 3 = 9 rectangles, or 9/20 of the original shape. Like this:

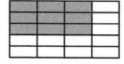

In each of these cases, you have ended up multiplying the numerators of the two fractions involved to find out how many little rectangles you want, and you have multiplied the denominators of the two fractions to find out how many little rectangles there are contained in the original shape. This process could be carried out to multiply any two fractions together, and is equivalent to 'just timesing the top two numbers and timesing the bottom two numbers.'

39. These squares have been divided into various different parts. Work out what fraction of the original square each part is.

a) In this problem A and B are half way along the side of the square.

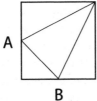

b) C and D are both ¼ of both away from the nearest corner.

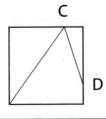

Right. We have covered addition and subtraction, as well as multiplication, so there is only one problem left to deal with and that is dividing fractions. This is what they told you to do at school: 'To divide fractions, you just turn the second one upside-down and multiply.' It doesn't sound like a rational thing to do at all, just a desperate strategy when faced with an unsolvable problem – when in doubt, move the numbers around on the page, write an equals sign, cross out some of the working, and boldly set down an answer at the end of it all.

There is an explanation for why this technique works. (If you really want it, you can find it in Appendix A at the back). But I don't like it, because it makes no use of rectangles, or pies, or marbles, or pencils, or anything else that mathematics teachers used to try and persuade you that what they were saying had some connection with the real world. There are no pictures to show what is going on, and it deals with strange fractions that are difficult to visualise. What is a '5/3/2/5' when it's at home? I wouldn't trust it. I wouldn't let my kids go trick-or-treating at its house. I wouldn't want to get too close to its musty smell and its blackened and broken teeth. It's the kind of thing you would see wearing plastic bags for shoes and muttering to itself as it walked in circles around the bus stop. Cross to the other side of the road is my advice. You see there is another way…

Take the problem: 2/3 ÷ 1/6. You have to think of this problem as 'how many 1/6s go into 2/3?'. The answer to this question is not at all obvious, since you have no way of comparing sixths with thirds. However, this is the same kind of problem that you came across when you were trying to add fractions, and if you use the same method that you used in that situation, you can represent 2/3 and 1/6 on two separate pieces of folded paper like this:

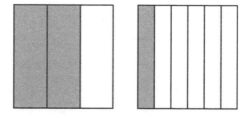

Then you fold the first piece of paper horizontally six times, and the second piece horizontally three times, just as before:

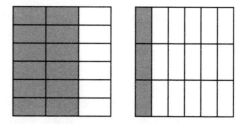

So the problem now is 'how many 3/18s go into 12/18?', and, if you think in terms of the shapes above, where 1/18 is represented by a small rectangle, this boils down to the question: 'How many sets of 3 rectangles make up 12 rectangles?'. The initial fraction problem has been converted into a problem involving only whole numbers, and the answer can easily be calculated as four. Once again, this technique applies to any division of fractions. Each time the aim will be to find equivalent fractions to the two fractions in the problem, both of which have the same bottom number. Once you have done that you can transform the original problem into the simpler problem of dividing two whole numbers.

10 WHAT IS THE (DECIMAL) POINT?

Decimals are a special way of writing fractions that are used when you want to apply our number system to numbers less than one. For the number 5472, the value of each digit depends on the column it is in. The two is in the units column, so it is worth 2×1 (or just 2), the seven is in the tens columns, and so it is worth 7×10 (or 70), the four is in the hundreds column, and so it is worth 4×100 (or 400), and the five is in the thousands column, and so it is worth 5×1000 (or 5000). The clever thing is, that as you move from right to left, the value of each column increases by a factor of ten. The next column after the 'hundreds' is the 'thousands', the next one after that is the 'ten thousands', and so on, and so on, and so on. All the rules we use for adding or subtracting or multiplying or dividing rely on this connection between the columns.

For example, if you want to add 347 and 291, you start off by counting the number of units, and writing down 8 in the 'units' column of your answer. Then you look at how many tens you have, and discover that you have thirteen. The rules say that you write down a three in the 'tens' column, and carry a one into the 'hundreds' column. But this relies on the connection between the values of the different columns. 'Thirteen tens' (130) is the same as 'one hundred' and 'three tens' (130). You have written down the three tens in the tens column, and carried the one hundred into the hundreds column. Then you count the number of hundreds, $(3 + 2 +$ the carried 1), and write down 6 in the hundreds column.

Decimals came about when people realised that it would be very useful if you could apply all the normal procedures for adding, subtracting, multiplying and dividing to fractions, rather than spend your whole time turning things upside down and inside out.

Some earlier civilisations made attempts to expand their number systems to deal with fractional numbers. The Babylonians were the most successful in their efforts. They were the ones who came up with a place-value counting system involving a mixture of base sixty and base ten, and they were perfectly happy to extend it to numbers less than one. They just invented new columns. So if you start at what for them was the '3600 column', the next column to the right would be the '60 column', the one after that would be the 'units column', the one after that would be the 'sixtieths column', and the one after that would be the '3600ths column'. As you move from left to right, each column has a value that is sixty-times smaller than the previous one. Further columns can be created if necessary by continuing to follow this pattern. However, the Babylonians never got around to inventing any equivalent for our decimal point, and so any number that they wrote down was ambiguous. It was not possible to tell what 'columns' each of its digits were in, except by referring to the context in which the number appeared.

Take the following Babylonian number (remember that < represents '10' and Y represents '1'): <<Y <YYYY <<<Y

It definitely translates as 21 14 31, but it is not possible to tell whether this means $(21 \times 60) + (14 \times 1) + (31 \times 1/60)$, or $(21 \times 1) + (14 \times 1/60) + (31 \times 1/3600)$, or $(21 \times 3600) + (14 \times 60) + (31 \times 1)$, or any other possible combination. And this is without worrying about the fact that, for a long time, they also had no symbol for zero, and therefore no way of showing that a column was empty. So, in fact, the number above could represent $(21 \times 3600) + (0 \times 60) + (14 \times 1) + (0 \times 1/60) + (31 \times 1/3600)$.

In terms of the number system that we use, the Arabs began to extend it to represent decimals, but it was not until the late 16th century that someone gave a full description of how such a system might work. The breakthrough came in a book written by the Belgian mathematician Simon Stevin (1548–1620), which was translated into English under the pithy title of '*DISME: The Art of Tenths, or, Decimall arithmetike, Teaching how to performe all Computations whatsoeuer, by whole Numbers without Fractions, by the foure Principles of Common Arithmeticke: namely, Addition, Substraction, Multiplication and Diuision.*'

Stevin stated that his aim was to teach everyone 'how to perform with an ease unheard of, all computations necessary between men by integers without fractions'. His book gave detailed explanations of how to add, subtract, multiply and divide any two decimal numbers. I don't know what he thought the problem was with computations between women – perhaps he was an unreconstructed sexist who thought that female computations didn't follow any logical pattern.

However, Stevin did not use the notation that we use nowadays. For example, if he wanted to express 364.957, he wrote 364 (0) 9 (1) 5 (2) 7 (3). 364 (0) represents 364 units (Stevin called this whole-number bit the 'commencement'), 9 (1) represents 9 tenths, 5 (2) represents 5 hundredths, and 7 (3) represents 7 thousandths. The (1) is short-hand for 'the first column for fractional numbers' (i.e. the 'tenths' column), the (2) stands for 'the second column for fractional numbers' (i.e. the 'hundredths' column), and the (3) means 'the third column for fractional numbers' (i.e. the 'thousandths column'). If necessary, this system could be extended to represent any decimal fraction you liked. Stevin called the (1), (2), (3) etc. 'signs'.

Later mathematicians invented all sorts of ways of writing decimal numbers. Eventually, the governments of different countries around the world had to decide which notation to choose. Unsurprisingly, they all had their own ideas on what was best for their country. If the world powers cannot decide what is the best way to separate whole numbers from fractional numbers, I don't hold out much hope for universal agreement on what is the best way to tackle global warming.

With the invention of printing, the easiest way to split a number into whole-number part and fractional-number part was to use the existing punctuation marks of the time, which were the comma (,) the full-stop (.), and the point (a full-stop in mid-air like this: ·). But nobody could decide between these three. The French went for the comma, because they already used the full-stop in printing to make Roman numerals more readable. However, English-speaking countries already used the comma to help them divide up large numbers (e.g. 123,456,678). It was the full-stop or the point for them.

But how to decide? The US did not beat around the bush. It decided on the full-stop and stuck to its guns. The British government, on the other hand, had a harder time making its mind up. It didn't want to offend anyone and so, of course, ended up offending everyone, and eventually having to do what the US had done in the first place. It first decided to use the point, because it was aware that some countries that had the misfortune of being outside the British Empire used the full-stop to write large numbers (e.g. 123.456.789), and it didn't want to confuse them. But then it turned out that the point was in common use by mathematicians to represent multiplication (e.g $3 \cdot 4 = 3 \times 4$), and so there was an international ban on its use in decimal numbers. With its tail between its legs, the British printing establishment gave in and adopted the full-stop in the late 19th century. It has been using it ever since.

The expansion of our number system to include decimal fractions revolutionized the way people went about calculating with numbers, especially once the French invented the metric system in the late 18th century. For example, the British system for measuring distances had all sorts of different conversions within it: 12 inches = 1 foot; 3 feet = 1 yard; 1760 yards = 1 mile. As a result, calculations were long and time-consuming. Once such systems were replaced by the far more logical metric system, where different units of measure were connected by factors of ten (10 millimetres = 1 centimetre, 100 centimetres = 1 metre, 1000 metres = 1 kilometre etc.) then calculations could be done far more quickly using the techniques that Stevin and others had developed for decimal numbers. The Australian Council of Educational Research estimated that the change to decimal money and metric measurement freed up at least eighteen months of mathematics lessons in primary schools, although Australia did not make this change until 1964. The US and France had been using decimal money for more than 150 years by then.

---

40. Write an addition sum that contains all the even digits, and an addition sum that contains all the odd digits, so that both sums have the same answer. You may use a zero as either an odd or an even digit, and you will need to use one decimal fraction and one improper fraction (a fraction where the numerator is greater than the denominator) to solve the problem.

---

11 MANIPULATING DECIMALS

Simon Stevin's trick was to expand the system of columns in the decimal place-value system. For any number, as you move from left to right, the value of each column decreases tenfold. So, if you start at the 'thousands' column, you move from 'thousands' to 'hundreds' to 'tens' to 'units'. Stevin showed that it was possible to continue this process. The column after 'units' becomes 'tenths', the column after that is 'hundredths', and so on. This makes it possible to express fractional numbers using the same 'column' system as is used for whole numbers, as long as you can express the fractional number in terms of tenths, hundredths, thousandths, and so on. And this, in turn, makes it possible to apply the same procedures as are used for whole numbers to fractional numbers.

In order to convert a normal fraction into a decimal fraction, it helps first to know how to convert a decimal fraction into a normal fraction. To convert 0.34, you have to remember the nature of the columns that each digit is in. In this case, 0.34 contains no units, three tenths and four hundredths. In other words 0.34 is $3/10 + 4/100$. Using addition of fractions, this simplifies to $34/100$.

In general (because the value of the columns as you move from right to left increases by a factor of ten) you can simply write the digits of the decimal number over the value of the smallest column that it uses. In the above problem, the digits are 34 (or 034 to be pedantic), and the smallest column used is the 'hundredths' column, and so the number is equivalent to $34/100$. In the same way, for the decimal number 2.578, the digits are

2578, the smallest column used is the 'thousandths', and so the number is equivalent to 2578/1000.

Converting from normal fractions is more complicated, and to do it you need to make use of the algorithm for long division. But now instead of being satisfied with a whole number 'remainder' at the end, you must continue the division.

For example, if you want to work out what 3/8 is as a decimal fraction, you set out the numbers just as for a long division:

$$8 \overline{\smash{\big)}\, 3}$$

Previously, you looked at this, said '8 goes into 3 zero times', shook your head and wrote down the answer as 0 remainder 3, but once decimal fractions come into play, it is possible to do better than that. The first step is to put in the decimal point (which secretly exists after the 3), and fill in some zeros. 3 is the same as 3.0, which is the same as 3.00, which is the same as 3.000, and so on, and so on. So:

$$8 \overline{\smash{\big)}\, 3.0000}$$

Now, you use the long division algorithm. As before, it continues to break up the division into helpful pieces. The first step here technically is:

$$\begin{array}{r} 0 \\ 8 \overline{\smash{\big)}\, 3.0000} \\ -\ \ 0 \\ \hline 3\,0 \end{array}$$

During this stage, you have discovered that, with 3 units, it is not possible to find an easy number of units that will divide between eight people, and so you have moved on to look at the tenths column:

```
        0.3
 8  3.0 0 0 0
 -  0
    ──
    3 0
 -  2 4
    ──
      6 0
```

In this stage, the long division algorithm has identified 2.4 as a number that is easy to divide between 8 people in terms of tenths. Each person gets 0.3 (or 3/10). It is easier to make sense of this, if you consider that 2.4 is 24/10 (which the algorithm prompts you to do by not including the decimal point between the 2 and the 4 in the working out). Once the 2.4 has been shared out, you are left with 0.6. Then:

```
        0.3 7
 8  3.0 0 0 0
 - 0
   ──
   3 0
 - 2 4
   ──
     6 0
 -   5 6
     ──
       4 0
```

The algorithm has this time identified that 0.56 (or 56/100) can be shared between eight people by giving each person 0.07 (or 7/100). After this 0.04 is left. Next

$$
\begin{array}{r}
0.3\,7\,5 \\
8\overline{)3.0\,0\,0\,0} \\
-\,0 \\
\hline
3\,0 \\
-\,2\,4 \\
\hline
6\,0 \\
-\ \ 5\,6 \\
\hline
4\,0 \\
-\ \ \ \ 4\,0 \\
\hline
0\,0
\end{array}
$$

Finally, the algorithm divides 0.040 (or 40/1000) between eight people, so that each person gets 0.008 (or 8/1000). It turns out that if three is divided between eight people, each person gets 0.375. Therefore, 3/8, as a decimal fraction, is 0.375.

Once again, the long division algorithm is simply a quick way of using repeated subtraction to solve the problem. The above process could just as easily be represented as follows:

$$
\begin{array}{rl}
\ \ \ \ 3.\,0\,0\,0 & \\
-\ \ \ 2.\,4\,0\,0 & \quad (0.3 \text{ lots of } 8) \\
\hline
0.\,6\,0\,0 & \\
-\ \ \ 0.\,5\,6\,0 & \quad (0.07 \text{ lots of } 8) \\
\hline
0.\,0\,4\,0 & \\
-\ \ \ 0.\,0\,4\,0 & \quad (0.005 \text{ lots of } 8) \\
\hline
0.\,0\,0\,0 &
\end{array}
$$

Therefore 3 divided by 8 is $(0.3 + 0.07 + 0.005) = 0.375$.

Not all fractions convert easily into decimal form. Some get caught in a never-ending loop. Take 1/3:

$$
\begin{array}{r}
0.3\ 3\ 3\ 3\ldots \\
3\ \overline{)\,1.0\ 0\ 0\ 0} \\
-\ \ 0 \\
\hline
1\ 0 \\
-\quad 9 \\
\hline
1\ 0 \\
-\qquad 9 \\
\hline
1\ 0 \\
-\qquad\ 9\ldots
\end{array}
$$

As soon as the remainder from one of the stages of the division is the same as a remainder you have had before, the whole process gets caught in a recurring cycle. 3 remainder 1, 3 remainder 1, 3 remainder 1 forever and ever and ever. In order to escape, you write that the decimal equivalent of 1/3 is 0.3̇, where the 'dot' means that the 3 keeps on repeating.

It is possible to find fractions that have more complicated cycles of repetition. 1/7 has a decimal equivalent of 0.14285714285714285... The digits 142857 repeat endlessly. In order to avoid imminent insanity, a dot is placed at either end of the repeating section. So, the decimal equivalent of 1/7 is 0.1̇42857̇).

---

41. Find the missing symbol in the following pattern:

Once you are happy with the ideas behind decimal fractions, you can add and subtract them using exactly the same procedure as for adding and subtracting whole numbers. If you want to add 12.34 and 2.5, you count four 'hundredths', eight 'tenths', four 'units' and one 'ten', and you write down the answer 14.84.

$$
\begin{array}{r}
1\,2.\ 3\,4 \\
+\quad 2.\ 5 \\
\hline
1\,4.\ 8\,4
\end{array}
$$

You have to be careful that you count the correct numbers from each column, and that is why you have to 'line up' the decimal points when you write down the sum. By doing this, you ensure that corresponding digits in the two different numbers are written beneath each other. The 'units' are underneath the 'units', the 'tens' are underneath the 'tens' and so on.

Similarly, when multiplying decimals, it is possible to use the procedure for long multiplication, just as for whole numbers. The only problem comes with deciding 'where to put the decimal point' at the end of the calculation. The rule for this is that you count the number of digits after the decimal point in the two numbers multiplied together, and make sure that there are the same number of digits after the decimal point in the answer. For example, in the sum 12.2×1.23, there are three numbers, in total, after the decimal point, and so, in the answer (which is 15.006), there must be three numbers after the decimal point.

The explanation for this rule is that a long multiplication is a series of simpler multiplications.

Here is the working for the example above:

$$
\begin{array}{r}
12.2 \\
\times \quad 1.23 \\
\hline
366 \\
2440 \\
+\ 12200 \\
\hline
15006
\end{array}
$$

The long multiplication has actually split the problem up into $(0.03 \times 12.2) + (0.2 \times 12.2) + (1 \times 12.2)$. The smallest multiplication involved here is the (0.03×12.2), in which 3 hundredths multiplies a number that contains, as its smallest 'part', 2 tenths. And 2 tenths multiplied by 3 hundredths is 6 thousandths. (You can check this by multiplying them together as fractions.) Therefore, you have to write down a 6 in the 'thousandths' column of the answer.

In the same way, if you had to multiply together 3.42 and 1.043, the smallest multiplication you have to do is to multiply 2 hundredths by 3 thousandths, and 2 hundredths multiplied by 3 thousandths is 6 one-hundred-thousandths. One-hundred-thousandths is the fifth column after the decimal point, and so that is where you write down the 6. The rule given above essentially works out the value of the smallest column that will be required to write down the answer, without going through the above process.

---

42. You are a travelling salesman in charge of persuading retailers to sell your deluxe collection of soups. Over a year, you clock up a total distance of 8400 miles, and the average price of petrol is £2.12 per gallon. Your car travels 32 miles to the gallon. How much does your company have to pay in petrol expenses?

You would have thought that there would have been some rule similar to the one for multiplying decimals to help you when you divide them, but instead we are asked to cheat, and use a trick to avoid ever having the misfortune of having to divide by a number with a dot (or a comma) in it.

At school, when faced with a problem involving division by a decimal fraction, you were told to move the decimal point as many places to the right as was necessary to make the number you are dividing by into a whole number. Then, you were ordered to move the decimal places the same number of places to the right in the number you were dividing.

So, if you were sat in front of the division $1.44 \div 1.2$, you moved the decimal place in the 1.2 one space to the right, to make it into 12, and then you did the same thing to the top number, to make it into 14.4. You were no longer faced with dividing by a decimal fraction, although you still had to deal with how to divide a decimal number by a whole number.

One thing at a time. What right do you have to go shifting decimal points around like this? The division above can be represented as a fraction (although not a very pleasant one):

$$\frac{1.44}{1.2}$$

You want to multiply this fraction by another fraction so that the denominator in your answer is a whole number, but without actually changing its value. In order to do this, you must multiply by a fraction that is equivalent to 1, because multiplying by 1 does not change the size of thing you are multiplying. Any fraction where the numerator is the same as the denominator is equivalent to 1. So 2/2, 55/55, 10/10, 0.3/0.3, and 11.14/11.14 are all exactly the same as 1.

In this case, in order to shift the decimal point one space to the right in the denominator you must multiply by 10, and so you multiply by the fraction 10/10. Then:

$$\frac{1.44}{1.2} \times \frac{10}{10} = \frac{14.4}{12}$$

Because 10/10 is the same as one, you have not changed the size of 1.44/1.2, but you have converted it into a division where the dividing number is an integer. In any division where you are dividing by a decimal fraction, you can use the same kind of trick. You multiply top and bottom of the division by whatever power of ten is necessary, and this has the effect of moving the decimal points in both numbers involved in the division a certain number of places to the right.

So for 1.2/0.34, the trick works like this:.

$$\frac{1.2}{0.34} = \frac{1.2}{0.34} \times \frac{100}{100} = \frac{120}{34}$$

In this case, you needed to move the decimal point two spaces to the right in order to make the dividing number into a whole number, and so you multiply top and bottom by 100. When you multiply the top number by 100, you have to fill in the gaps left by the moving decimal point with zeros. The decimal point has not disappeared. 120 is the same as 120.0 – the decimal point in 1.2 has moved two spaces to the right.

That still leaves you with the problem of having to divide a decimal number by a whole number. In the first example above, even after you used the trick, you were left with the division:

$$12 \overline{)\,14.4}$$

Fortunately, such a division can be dealt with in the same way as you dealt with the division of 3.0000 by eight earlier. The long division algorithm works on it in exactly the same manner:

```
          1.2
    1 2 | 1 4.4
    -     1 2
          ___
           2 4
    -      2 4
          ___
           0 0
```

It turns out that '1.44 divided by 1.2' is the same as '14.4 divided by 12', which is 1.2.

> 43. You are still a travelling soup salesman, although there are rumours that you are soon to be promoted. It has been a while since you bought flowers for your wife, and so you wander into a florist to see what is on offer. You can buy twelve roses for £7.68, or you can buy five for £3.45. Which of these is the best value for money?

Charlie is only partially aware that Mr Barton is talking about how to divide decimal numbers. The monotonous voice of his teacher has become an almost-comforting backdrop to his wandering thoughts, which revolve lazily about the issue of whether his best friend really did get that scar on his face as a result of being sat on by a horse at an early age. The bell goes, and Charlie's eyes refocus. It is time for lunch, and he needs all his wits about him, if he is to make sure of a sufficient supply of carbohydrates to propel him through the afternoon.

Mr Barton dismisses the class, and locks the door of the classroom from the inside, so he can remove his cellophane-wrapped sandwiches from their Tupperware prison. He chews his way through the bland fillings with the slow and mechanical action of a cow chewing the cud, whilst trying to filter out the noise of the children streaming past his door on their way to the dinner hall.

Charlie temporarily forgets about the special laws governing his movements. He joins the queue at the doors of the dining hall, and manages to stealthily slip up the outside of it, without the on-duty teacher noticing. Once inside, he untucks his shirt and loosens his tie, bypassing the salad counter with a contemptuous glance at Bernadette as she fills her plate full of lettuce. His eyes are on the dull and battered metal containers on the hot food counter. Taking a tray, he piles it high with hamburgers, fries, sausage rolls, and two cokes, before sitting down at a table with his friends. There is little time for conversation, as they each work their way through the mounds of fat and sugar in front of them. Charlie can already feel the reassuring buzz of a glucose high coming on. It is going to be a good afternoon.

12 ONE HUNDRED PERCENT

Percentages are one part of school mathematics that do actually turn up on a regular basis in daily life. They form part of the constant barrage of information that the financial world throws at us. Billboards promoting loans, banks advertising interest rates, stock markets charting growth, shops touting sales – all make use of percentages as a method of conveying their messages. It is one of the few topics for which Mr Barton can genuinely feel justified in answering Charlie's umpteenth request to know the relevance of what he is learning to his everyday existence with the stock answer of: 'It will help you when you are older'.

Percentages are a slightly different kind of scale that we use to measure things, but, where fractions label a pizza as 'one pizza', percentages call a pizza '100% of a pizza'. This might seem like overcomplicating things a bit, but it does have the advantage that, if you start cutting up the pizza, you can still talk about the pieces in terms of whole numbers. So half a pizza is 50% of a pizza, a quarter is 25%, and a fifth is 20%.

Because percentages are a scale used for measuring different quantities, problems involving percentages can be solved using proportional thinking. For example, if you buy a packet of plain digestive biscuits for £1.20 excluding VAT, and you want to have the correct amount of cash ready at the till, you have a couple of choices. You can either put the biscuits back on the shelf and purchase some chocolate digestive biscuits instead (because these are strangely not subject to VAT), or you can work out what 17½ % of 120 is.

If you plump for the second option, it is possible to solve the problem in the same way that other proportional problems can be solved. In this case, 120 pence is the 'whole', and you want to find out what 17½ % of it is. But, using percentages, the 'whole' is labelled as 100% and so, for the purposes of buying this particular packet of digestives, 100 % is equivalent to 120 pence. Once you have made this connection, you can use the techniques of proportional thinking.

Percentage	Price (pence)
100 %	120
1 %	1.20
17 1/2 %	17 1/2 × 1.20 = 21 pence

The first step is to find 1% of the original amount, and then it is possible to find any other percentage by multiplying 1% by the specified number. In this case, you wanted 17½ %, and so you multiplied by 17½, but if you had wanted 46%, you could have simply multiplied 1.20 by forty-six.

The full price of the packet of biscuits is 120 pence, plus 21 pence for VAT. In total, you need to hand over 141 pence. There is a slightly quicker way to reach this conclusion (aside from asking the shopkeeper to mark all his products with the full price). It relies on the following argument. If the basic price of the biscuits is 100%, and VAT adds a further 17½ % to this, then the final price will be 117½ % (100 + 17½) of the basic price. Armed with this information, you can calculate the final price immediately:

Percentage	Price (pence)
100 %	120
1 %	1.20
117 1/2 %	117 1/2 × 1.20 = 141 pence

If you consider the calculation in this way, it is not necessary to do the final addition, where you combine the initial price with the tax on it.

As an alternative to both of these techniques, once you have established that 100% is equivalent to 120 pence for VAT, you may prefer the quick method of working out 17½%, shown in the table below:

Percentage	Price (pence)
100 %	120
10 %	12
5%	6
2 1/2 %	3
17 1/2 %	21

It is generally fairly easy to work out what 10% of an amount is. From this, it is possible to find what 5% and 2 ½% is by halving the answer for 10%. And then you can calculate 17½% of the total by adding together your results for 10%, 5% and 2 ½ %.

The same ideas can be used to calculate what percentage something is of another thing. Say that it had come to your attention that of the forty socks you owned in this world, only twenty-one of them were undamaged, and you want to work out what percentage of your socks are undamaged. In this example, the forty socks that you own is the 'whole', and so it is equivalent to 100%. Once again, once you have made this link, the rest of the problem can be solved in a similar way to previous proportional problems:

Percentage	Number of socks
100	40
100 ÷ 40 = 2.5	1
21 × 2.5 = 52.5	21

This problem is different from the previous one in that the first step is to find how many percent one sock is. Once this result has been calculated – in this case one sock represents 2.5% of your earthly allotment of footwear – it is possible to calculate what percentage any number of socks represents. For this example, you wanted to know what percentage twenty-one socks was out of a total of forty, and so you multiplied by twenty-one, but if you had wanted to know what percentage twelve socks was, you would multiply by twelve.

---

44. If seventy percent of the people in the room have a wooden leg, seventy-five percent of people in the room have parrots on their shoulders, and there is nobody in the room who does not either have a wooden leg or a parrot, what percentage of people must have a wooden leg and a parrot?

Of course, Mr Barton can make things more difficult. He has further tricks up his sleeve. He can write a problem like this on the board: 'If a pair of jeans in a sale costs £17.50 after a 20% decrease, what was their original price?' You could answer that the original price of the trousers was of little interest to you. But if you want to play Mr Barton at his own game, you will have to do better than that.

The trouble is that in this situation you are not given the value of the 'whole', which in this case is the original cost of the trousers. All you know is that, after a 20% reduction, they cost £17.50, and this is equivalent to saying that 80% of the original price is £17.50. In order to solve the problem, you need to use this fact to find out what 100% of the original price is – because 100% of the original price is the original price itself – and this can be done by using a similar method to the one in the problem above. You first calculate 1%, and from this find 100%.

Percentage	Price (pence)
80	1750
1	$1750 \div 80 = 21.875$
100	$21.875 \times 100 = 2187$

You can use exactly the same procedure to work out the initial price after a percentage increase. For example, you buy a packet of biscuits for £1.20 including VAT (they are not chocolate-coated), and you want to find their price without the tax. VAT is 17½%, and so the price of the biscuits now is the initial price after a 17½% increase. In other words, if the starting price is 100%, the price with VAT is 117½%. And so:

Percentage	Price (pence)
117.5	120
1	$120 \div 117.5 = 1.021$
100	$1.021 \times 100 = 102$ (to the nearest pence)

For each of the procedures mentioned above, there is a 'rule' which you can obey in order to reach the answer. To find x% of an amount A, you multiply A by x and divide by 100. To find what percentage an amount C of another amount D is, you divide C by D and multiply by 100. To find the initial price of an item, given its price, E, after a percentage increase (or decrease – where a decrease is represented by a negative number) of y%, you divide F by $(100 + y)$ and multiply by 100. But all these rules hide the fact that, in each case, the problems are solved by reference to simple proportion, so that similar techniques can be applied to all of them.

---

45. After years on the waiting list, you finally receive your membership for an exclusive golf-club which does not allow women in the clubhouse. But the annual fee has gone up by 20% to $1450, because the ruling on women has been changed, and it is necessary to build changing-rooms for them. What was the fee for last year?

13 SOMETHING OF INTEREST

Mathematics and money go hand in hand. People have been doing business since the dawn of time, and where there is business, there are numbers. Perhaps the most common appearance of percentages is in their use to express rates of interest on accounts and loans. In this context, it is necessary to deal with compound interest, where a particular rate is continuously applied to the sum of money concerned over a period of time.

I say that it is necessary to deal with interest, but that isn't strictly true. You can, in fact, refuse to have anything to do with it on religious grounds. People have been making loans and charging interest on them (a practice known as usury) for 4000 years or more, and for the same amount of time other people have been shaking their heads in distaste, although the exact definition of usury has changed from place to place.

The earliest references to usury are found in religious manuscripts of India, and date back to around 2000-1400 BC. In these references, anyone charging any form of interest is a usurer, and despised for his calculating ways. At this time, it was forbidden for members of the higher castes to be involved with such devious practices. Later, 'usury' referred only to the charging of excessive interest on loans.

The Old Testament has plenty to say about usury. The Hebrew word for interest is '*neshekh*', which literally means 'bite'. The Books of Exodus and Leviticus both lay down that it is prohibited to charge excessive interest on loans to the poor and the sick, whilst Deuteronomy bans all loans of this sort, unless they are to foreigners.

The Christian Church followed the Jews in its distaste for usury, continuing in the spirit of Jesus' notorious expulsion of the money-lenders from the temple. The Catholic Church first banned the clergy from charging interest in the 4th century AD, and then, in the 5th century, extended this ban to all those who followed the faith. In the 7th century, Pope Leo XIII branded usury as 'a demon condemned by the Church but practised in a deceitful way by avaricious men' in his *Rerum Novarum*. However, with the advent of Protestantism came a more relaxed attitude towards it. Charging interest was accepted as a necessary part of financial transactions, which only became sinful if the interest charged was excessive. In 1987, when Pope John Paul II published a tract on his views on current affairs entitled *Solicitude Rei Socialis*, there was no mention of usury at all, except as a recognition of the Third World Debt crisis. (In 1980, Developing Countries had a debt of $567 billion. Since then, they have paid $3450 billion in interest and repayments, but despite this, their collective debt has risen to $2070 billion.)

In contrast to Christianity, some sections of the Muslim world have persevered in their refusal to charge interest on loans. This position arises from statements made by the prophet Muhammad in the Qur'an around 600 AD, in which he referred to the practice as '*riba*', which literally means 'excess'. As a result, Islam has developed a separate system of banking, which remains true to the principles that Muhammad laid down. In this system, the charging of interest is not allowed. The lender is forbidden from deriving any benefit out of making a loan, and, on the strictest interpretations of this demand, this includes indirect benefits, such as accepting a meal or some small token of gratitude from the person to whom he has lent his money. If only modern day politicians could abide by the same principles.

In fact, Islamic banks run under these laws are all-round decent places. If they lend money to an entrepreneur, they share an equal burden of responsibility for the use to which the money is put. If the venture fails, they must share in the losses that it incurs. In fact, according to Islamic scholars, when the bank makes a loan, the only benefit it can expect to gain is to receive the blessing of Allah. The idea is that the purpose of a loan is solely to benefit the community to which the money is lent. In contrast to this, I feel very strongly that it is of no benefit to the community to charge me £25 for exceeding my overdraft limit for just one day. I don't think my banker is getting any respect from Allah for that. In my opinion, he needs to spend more time considering his spiritual side.

> 46. You are unable to get over the injustice of your position. Why should you be made to pay for the building of women's changing-rooms in your golf-club? You decide to do a survey of all the male members to see what their views on the subject are, and discover that, of the 420 male members, 305 of them are in favour of banishing female golfers to an outhouse behind the notorious bunkers on the far side of the 13th green. What percentage of the total membership of the golf-club is in support of this scheme, if 120 female members joined at the start of the year?

However, if you don't want to avoid interest on religious grounds, then it will probably form an important part of your financial life. So it is worth having a look at it. Imagine that you take out a loan for $200, and that the rate of interest is 8% per year. In this situation, the 'whole' or 100% is the initial loan. After one year, you will owe the initial loan (100%), and the interest for one year (8%). In other words, you need to calculate what 108 % of $200 is. Using the methods from the previous chapter:

Percentage	Money (dollars)
100	200
1	$200 \div 100 = 2$
108	$108 \times 2 = 216$

So, if you don't pay any of the loan back, at the end of the first year your debt is $216 in total.

To calculate how much you will owe after the second year, it is necessary to consider that you will be charged interest for this year on all of what you owe. In other words, $216 is now the 'whole' or '100%' on which you are going to be charged 8% interest. You need to work out what 108% of $216:

Percentage	Money (dollars)
100	216
1	$216 \div 100 = 2.16$
108	$108 \times 2.16 = 233.28$

If you continue to ignore the threatening letters from the loan company demanding repayment, then at the end of the second year, you will owe $233.28.

And if you remain unconcerned by your descent into a never-ending downward spiral of debt, then after a third year it will be necessary for the loan company to send in their thugs to demand $251.94 (or goods of an equivalent value), since this is what 108% of $233.28 turns out to be. At the beginning of each year, the '100%' from which the 8% interest is calculated changes to become the total debt from the previous year.

It is possible to summarise the calculations that you have done to find your debt after each year. In each case, you are dividing the debt from the previous year by 100, and multiplying by 108. This is equivalent to simply multiplying by 1.08 (108 ÷ 100).

Time (years)	Money owed (dollars)
0	200
1	$1.08 \times 200 = 216$
2	$1.08 \times (1.08 \times 200) = 1.08 \times 216 = 233.28$
3	$1.08 \times [1.08 \times (1.08 \times 200)] = 1.08 \times 233.28 = 251.94$

From this table, it is possible to see that to find the amount of money owed after x number of years, it is necessary to multiply the initial amount borrowed by 1.08^x. In this particular example, after two years, you multiply 200 by two 1.08s (or 1.08^2). After three years you multiply 200 by three 1.08s (or 1.08^3).

To find the amount of money you would owe after ten years, if you make no repayments during that time, you multiply 200 by ten 1.08s (or 1.08^{10}). Your calculation would be 200×1.08^{10}, and your total debt would be around 430 dollars.

This method can be generalised for any rate of interest for any sum of money borrowed (or invested) over any number of years. If S is the sum of money borrowed (or invested), x is the rate of interest (expressed as a decimal number), and n is the number of years for which you have taken the loan (or invested your money), then the amount of money you owe can be found by using the following formula: $S(1 + x)^n$. [The $(1 + x)$ comes about because you are adding the interest charged to the initial sum borrowed. In the above problem, you found that every year you owed 100% + 8%. In the general case, you owe 100% + x% (or $(100 + x)$%), and when it comes to calculating your debt, this is equivalent to multiplying by $(1 + x)$.]

So if you had taken out a loan of £500 at a rate of 12% over a period of 4 years, then at the end of this period, you would owe $500(1 + 0.12)^4 = 500 \times 1.12^4 = £786.76$.

---

47. A rectangle has length 10 cm and width 4 cm. If the length of the rectangle is increased by 20%, by what percentage must the width of the rectangle be decreased by?

---

14 PRUDENCE IS A VIRTUE

I normally throw away those pamphlets that your bank sends you with a picture of a very sensible, yet quite attractive, woman on the front, and a whole load of advice inside them about how to manage your finances better. But they aren't just an attempt to drum up business for the bank. There are definite reasons why some of the advice is worth taking.

'IT IS IMPORTANT TO PLAN FOR THE FUTURE – START SAVING EARLY'

An argument to back up this piece of advice comes from looking at the way compound interest works. There is a rule of thumb that helps you to work out how much time it takes for your money to double when you invest it at a particular rate. The rule states that, at an interest rate of x%, your money double in $70/x$ years. So at a rate of 7%, you double your money in 10 years, and, at a rate of 2%, you double your money in 35 years.

From this, you can see that the longer you leave your money in a savings account, the faster the rate at which your savings grow. If you invest $100 at a rate of 7%, then, after ten years, you investment will double to $200, an increase of $100. However, over the next ten years, the $200 will double to $400 (an increase of $200), and over another ten years, the $400 will double to $800 (an increase of $400). After ten years, your initial investment doubles, after twenty years it quadruples, and after thirty years it octuples. In order to cash in on the larger increases that happen further down the road, you need to get your money into a high-interest savings account fast.

---

48. You reach the counter in your local store clutching a floral-print dress with a price tag of £64, only to find yourself bewildered by the variety of special offers that are available to you. The assistant says that you are eligible for a 10% discount because you are paying by cash, a 15% discount because you are a long-standing customer, and a 20% discount because there is a sale on. In which order should you take these discounts in order to get the best price?

'IT IS A GOOD IDEA TO BE REGULAR – TRY TO PUT AWAY SMALL SUMS REGULARLY, RATHER THAN LARGE SUMS OCCASIONALLY'

This piece of advice applies to many areas of life, most particularly eating and drinking, but I will focus here on how it might affect your savings. If you invest £1000 every year for ten years into an account that pays 5% interest, then at the end of the ten years you will have £13 207 (to the nearest pound). The calculation to arrive at this figure is difficult and lengthy, but it is really just a complicated version of the compound interest problems that you dealt with earlier.

It might help to imagine that each year you open a new savings account, and put £1000 in it. Then, the £1000 you invest at the start of the first year will sit in its account and gain interest for ten years, whereas the £1000 you invest at the start of the second year will gain interest for only nine years, and the £1000 you invest at the start of the third year will be in the account for only eight years, and so on, and so on, up to the £1000 you invest at the beginning of the tenth year, which will gain only one interest payment.

The problem then is to calculate how much money is in each of these ten different accounts at the end of the tenth year, and you can speed this up by using the formula for compound interest that you derived earlier:

Start value of investment	No. of years invested	Final value of investment
£1000	10	$1000 \times 1.05^{10} = 1628.89$
£1000	9	$1000 \times 1.05^9 = 1551.33$
£1000	8	$1000 \times 1.05^8 = 1477.46$
£1000	7	$1000 \times 1.05^7 = 1407.10$

At the end of the tenth year, the account opened at the start of the first year contains £1628.89, the account opened at the start of the second year contains £1551.33, the account opened at the start of the third year contains £1477.46, and so on. Similar calculations can be made for the accounts opened at the start of later years. Once these have been done, it is possible to find the total sum of money in all the accounts by adding the results. You should find that it comes to £13 207.

This technique can be applied to any similar system of investment, and, in fact, it is possible to derive a formula for the total amount of money in such a system after any period of time (although the proof of the formula is complicated). If S is the sum invested yearly, x is the rate of interest, and n the number of years for which you have been making the payments, then the amount of money that you have at the end of n years is given by:

$$\frac{S(1 + x)[(1 + x)^n - 1]}{x.}$$

You can use this formula to check that the result obtained above is correct. Just to show what a good idea it is to invest regularly, it is worth comparing the above system with a large one-off payment. In your system, the maximum amount of money that you invested was £10 000, and this amount was only invested for one year (the tenth year), at the end of which your total investment was worth £13 207. However, if you just decide to invest £10 000 in one go, rather than in stages, it is surprising how long it takes before you make as much money as this. In fact, it takes six years for a one-off investment of £10 000 to generate the same amount of income. You can check this by using the formula for simple compound interest: $10\ 000 \times 1.05^6 = £13\ 401$.

> 49. A newspaper article announces that a man involved in a shark attack lost 15% of the length of his leg, which now measures 70cm. What was the initial length of his leg?

So both regularity and forward-planning lead to a healthy financial situation. And if you combine the two, you can laugh all the way to the bank. Just to illustrate this, imagine that for ten years you had stuck to your system of paying in £1000 yearly to your savings account. At the end of this time, satisfied with the healthy-looking figures at the bottom of your bank statement, you abandon your yearly payments. You are happy just to receive the interest payments on the amount you have managed to save so far. In other words, from now on, you are simply receiving compound interest of 5% on an original investment of £13 207 (the sum that you have saved over the ten years of your regular payments of £1000).

Unknown to you, your new flatmate has a habit of going through your correspondence (as well as stealing into your room at night and watching you sleep), and she is aware of your growing fortune. She decides to follow your lead, and invest £1000 yearly into a similar savings account. She will not rest easy until she has as much money as you, and made up for her ten years of wasted investment opportunities.

The table below shows how much each of you have invested in your respective bank accounts after different periods of time.

Number of years later	Value of your investment (to nearest pound)	Value of flatmate's investment (to nearest pound)
0	13 207	1000
10	21 512	13 207
20	35 042	34 719
21	36 794	37 505

The amount of money in your account can be calculated using the formula for simple compound interest. Your initial investment over this period of time was £13 207, and so its value after twenty years is given by 13 207 × 1.05²⁰, which comes to £35 042. The amount of money in your flatmate's account can be calculated using the complicated formula for investment schemes where payments are made yearly into an account. So after twenty years, the value of her investment is

$$\frac{1000 \times 1.05 \times (1.05^{20} - 1)}{0.05} = £34\ 719.$$

Overall, your flatmate has had to invest £21 000 in order to match your funds, whilst you have only invested £10 000. It has taken her twenty-one years of obsessive saving to catch up with you financially – twenty-one years in which daily she waited for her bank statement with a mixture of excitement and fear, twenty-one years in which she was forced to humiliate herself by trawling through your dustbin in search of receipts. In short, twenty-one years of her life have been wasted, because she, unlike you was neither forward-thinking, nor regular. You ought to ring up the bank and get the number of that sensible, yet attractive woman from the pamphlet cover. You are well-suited to one another.

---

50. In a mathematics exam, there were two questions. Number 1 was solved by 70% of the students, and Number 2 was solved by 60% of the students. Every pupil solved at least one, and 9 pupils solved both. How many pupils took the exam?

---

15 TWO HUNDRED PERCENT

It doesn't seem like a bad idea to come up with a system that avoids the use of fractions, but it does give rise to some strange results. It might not worry you that a percentage can be a number larger than 100. If the energy that a soccer player puts into a game is measured to be the equivalent of thirty-two bars of chocolate, then an energy output of 150% would be the equivalent of forty-eight bars of chocolate, an energy output of 200% would be the equivalent of sixty-four bars of chocolate, and an energy output of 325% would be the equivalent of 104 bars of chocolate. Once the percentage scale has been applied, so that 100% represents an energy output of thirty-two pints of lager, it is possible to extend it as far as you like.

However, it is always a bit of a surprise to learn that if you increase the price of something by 200%, it triples. Surely it should double?

---

51. As a result of your inflammatory survey, a group of female golfers decide to take justice into their own hands, and take to attacking male golfers with putters and sand irons. The committee introduces armour-plated golf buggies to deal with this new threat, but is forced to advertise that the following year's membership fee will increase by 5% (from $1450) to cover their costs. However, after the capture of the renegade ladies, the committee no longer feels that it is necessary to raise membership fees, and announces that its member should reduce the advertised fee by 5% to find the actual fee for the following year. How much will members have to pay next year?

If you take a pair of trousers that cost £17.50, and decide to increase their price by 200%, the first thing to do is to calculate what 200% of the price is using THE METHOD:

Percentage	Price (pence)
100%	1750
200%	2 × 1750 = 3500

But this is just the increase in the price. To find the new price you have to add it to the old price, which means that the trousers now cost 1750 + 3500 = 5250 pence. By adding 200% to the original price, you have effectively added twice the original price to itself, which means that the new price is three times the original.

In the same way, if you increase the price by 300%, you end up with quadruple the initial price, because you have added 300% to the initial 100%:

Percentage	Price (pence)
100%	1750
300%	3 × 1750 = 5250

After a price increase of 300%, the jeans cost 1750 + 5250 = 7000 pence (which is four times more expensive than 1750 pence).

There is stranger still to come. You would have thought that if you take the same pair of jeans, increase their price by 10%, and then decrease it again by 10%, you would end up with a pair of jeans costing £17.50. But you don't. Which is odd.

The reason for this bizarre outcome is that there are two stages involved in this problem, and the 'whole' (or the 100%) is not the same for both. In the first stage, you have to increase the price of the jeans by 10%. So:

Percentage	Price (pence)
100%	1750
1%	1750 ÷ 100 = 17.50
10%	10 × 17.50 = 175

Since 10% of the original price is 175 pence, an increase of 10% means that the trousers now cost 1750 + 175 pence, which is 1925 pence.

In the second stage of the problem, you have to decrease this new price by 10%, but this means that you are now treating the new price of 1925 pence as 100%, and not the original price of 1750 pence. 10% of the new price is going to be different from 10% of the old price, and so it is not surprising that after the 10% decrease, the cost of the trousers is different from the original cost. In fact:

Percentage	Price (pence)
100%	1925
1%	1925 ÷ 100 = 19.25
10%	10 × 19.25 = 192.5

And so a 10% decrease will mean that the final price of the trousers is 1925 − 192.5 = 1732.5 pence, which is slightly less than the original price.

---

52. At the Kigali Christmas Expo, I ate four plates of fried green bananas, and bought ten pounds of dried beans. When I weighed myself at the end of my visit, my weight had increased by 10%. I quickly went on a diet to return to my former sleek figure, but had to return to the Expo a second time, where I ate twice as many plates of bananas, and bought ten pounds of maize flour. I weighed myself again, and found that my weight had increased by 11%. What was my original weight?

PART THREE

FEAR OF THE UNKNOWN

1 ALGEBRA AND BROKEN BONES

Charlie is doing things his own way. He believes in self-expression, and he doesn't understand why Mr Barton is always trying to change the way he approaches things. Take algebra, for example. Algebra is just another word for finding the missing number. Charlie has sat and listened to Mr Barton's explanation about 'what you do to one side, you have to do to the other', but it all sounds a bit communist for his liking, and he has always been able to solve problems by his own special methods.

So, after disturbing the boy in front of him for several enjoyable minutes by gently kicking the bottom of his chair, he settles himself down to deal with the problems Mr Barton is writing on the board with a general feeling of confidence. $2x + 11 = 21$. No worries – the $2x$ must be the same as 10 because $10 + 11 = 21$, and so x must be equal to 5. Charlie gives himself a tick, because he is pretty sure that Mr Barton will disagree with how he has gone about this problem. He moves on to the next question, $4y - 9 = 19$. Once again, Charlie's method works flawlessly: $y = 7$. He celebrates by stealing the rubber off the desk next to him. Next, $3t - 5 = 2t + 1$. Oh dear...

Charlie feels a dark, sinking feeling deep inside him. Nothing is certain any more. How does he deal with the madness that he is suddenly faced with? There are two 't's in this problem. What do you do when there are two 't's? He whispers his question to his next-door neighbours, but he has already alienated them by kicking their chair and stealing their rubber respectively. He surrenders completely to a feeling of total helplessness, and lets his

MATHEMATICS **MINUS** FEAR

head slowly sink into his hands. Once again, the dark forces of mathematics have combined to defeat him. It won't be long before Mr Barton senses this, and comes to mock him in his defenceless state. He can hear the squeaking of dirty brown brogues now.

At school level, algebra is the study of generalised number. It can range from being incredibly abstract to reasonably concrete. If you write down the symbol 'x' and nothing else, that symbol stands for 'any number at all'. It is as if, were you to look underneath the symbol, you would find a box that contains all of the infinite numbers in existence. Slightly less abstract, but definitely not specific, is an expression like $2x$. In this case, $2x$ stands for 'two times any of the numbers that exist', which, in fact, means that it also stands for any number at all, since every number is twice another number. Hmmm...

Things become a little clearer, when you start attaching such expressions to particular situations. A middle-aged woman refuses to tell you her age, and so you can refer to her age as being x (although in this case, the circumstances of the situation limit the possibilities for x – she can't be a negative age, and she can't be older than 150). She also informs you that her son is thirty years younger than her. This doesn't help you to work out the age of her son, but it does mean that you can refer to his age as being $(x - 30)$. Finally, she lets slip that the number on a passing bus is three times the age she will be in five years time. Again, this is no help at all if you want to know the number on the bus, but it does allow you to express it in terms of x. In five years time, the woman will be $(x + 5)$ years old. The number on the bus is three times this, and so it can be expressed as $[3 \times (x + 5)]$.

All of these are examples of algebraic expressions. In each case, the unknown (or variable) potentially stands for all the numbers in existence. If you want to get more particular, you need more information. If you get more information, you might be able to form some equations. To form an equation you need to know that two different things have the same value. For example, if the lady admits that her son is 20, you know that $(x - 30)$ is equivalent to 20, and you can write down the equation $x - 30 =$

20. This equation has only one solution, 50.

Not all equations are as specific. It is possible to come up with equations that have as many solutions as you like. For example, $x^2 = 16$ has two solutions (4 and -4), $(x - 1)(x - 2)(x - 3) = 0$ has three solutions (1, 2 and 3), and $x = x$ has an infinite number of solutions.

---

53. The swimming pool at the hotel in Akagera Game Park needs to be topped up, but the plumbing has once again been destroyed by a rogue elephant. The staff are able to work together to fill it at a rate of 20 litres per minute. If there were 2000 litres of water in the pool at the start of the operation, write and expression for the amount of water after m minutes.

So far, I have mentioned expressions and equations that only contain one unknown, but it is perfectly possible to come up with equations and expressions that contain two or more unknowns. The lady can tell you that the number of ice creams that she has eaten in the course of her life is equal to the sum of her age and her husband's age. Then if you call her age x, and her husband's age y, the number of ice creams that she has eaten is the expression $x + y$. If she goes on to say that the number of ice creams she has eaten is 80, you can write down the equation $x + y = 80$. This equation is called indeterminate because there are an infinite number of possibilities for x and y. The lady could be 40 and her husband could be 40, or they could be 41 and 39, or 100 and -20, or 79½ and ½, although some of these are physically impossible and some of them are illegal.

In order to fix the values for x and y, you need more information. For example, if she also tells you that she is older than her husband, and their difference in age is 20, you can write down the equation $x - y = 20$. You now have two equations that must be true at the same time. Such equations are called simultaneous equations. In this case, you now have enough information to find the value of x and y, but this is not always true of simultaneous equations. They can still be indeterminate. In general, if you have two variables, you will need at least two equations connecting them, if you have three variables, you will need at least three equations connecting them, and so on, and so on.

Once you have come up with the idea of representing an unknown number by a symbol, the next difficulty that faces you is how to deal with them. The history of elementary algebra consists of the various inventions and improvements that different cultures have come up with to deal with algebraic expressions and equations.

Some techniques have been discarded, as better ones have been found, and there are a couple of examples of these in the chapters to come. The techniques that have survived form the basis of what you were taught at school. I am sure you will remember instructions like 'do the same to both sides', and 'cross-multiply', and 'expand the bracket by multiplying everything inside it'. It is these 'laws' of algebra that are the problem. They form a series of logical steps which lead you towards a solution of an equation or a simplification of an expression – but it is often not at all clear where the logic lies. They often feel like a series of commands from an unsympathetic despot, whose one aim is to force you to fill pages of paper with strange symbols according to random rules. But I hope that in what follows, it will become clear that all the rules of algebra, both those that have fallen out of use and those that Mr Barton continues to drill into his students, have their roots in an understanding of how numbers in general work.

The notion of pain is embedded in the word 'algebra' itself. 'Algebra' is an 'Englishification' of the Arab word '*al-jabr*', which al-Khwarizmi used as a technical term in connection to solving equations. It describes the process of taking a term from one side of an equation to the other. For example, if you have the equation $4x = 2 - 2x$, you can 'take the $2x$ to the other side', and write $4x + 2x = 2$. However, al-jabr can also describe the process of setting a broken bone. There are probably many people in the world who feel that the two operations are equally gruesome.

In Mediaeval Spain, due to the influence of the Moors, a barber called himself an *algebrista*, because he generally did a bit of bone-setting and blood-letting on the side to supplement the income he received from styling hair. That is why the traditional sign for a barber is a red-and-white striped pole – the red and white symbolise blood and bones.

Rumour has it that Stephen Hawking was instructed by his publishers not to include an equation anywhere in the first half of a scientific book that he intended for the general public. It was apparently felt that the sight of an equation would immediately cause the average reader to put the book back on the shelf and walk hurriedly away.

---

54. Think of a number. Add four, and double your answer. Then subtract eight, and divide by two. Your answer will always be the number you started with. How does this trick work?

---

2 DOING THE SAME TO BOTH SIDES

Some of the earliest evidence we have of people attempting to go about the business of solving equations involves the ancient Egyptians. Equations for them were, in fact, word problems, in which the aim was to find some unknown number. This was because they had not got around to using symbols like x or y, and also, they did not feel the need to follow a particular method, but picked whatever approach they felt to be useful at the time.

The Egyptians' favourite method was refreshingly haphazard. It is called 'The Method of False Assumption', and has been very popular ever since they introduced it. In fact, it was still common in Europe until one hundred years ago, and I am sure it lives on in the minds of independent spirits like Charlie.

Here is Problem 26 from the Rhind Papyrus – the one that Mr Rhind found in a bazaar in Egypt. The scribe has made no attempt to make it interesting by pretending we are trying to look for the unknown length of a rabbit, or something equally ridiculous. Rabbits aren't going to make the problem any easier. 'A quantity and its quarter added become fifteen. What is the quantity?'

Unsurprisingly, the Method of False Assumption relies on making a false assumption. In this case, the scribe guesses that the quantity is four, because it is easy to work out what four plus a quarter of four is. It is five.

He then points out that, in the problem, the answer to this sum should be fifteen, which is three times greater than five.

He therefore adjusts his false assumption by a factor of three, and reaches the true conclusion, which is that the quantity is twelve.

It is good to find people being so casual about solving equations, but this approach only works for linear equations in one variable (equations that contain just one unknown, and in which the unknown is not squared or cubed or anything like that – examples are: $3x + 4 = 2$ and $2x - 1 = x + 4$). As soon as you start dealing with things containing higher powers (e.g. x^2) or containing more than one variable (e.g. x and y) you have to find another way. False assumptions can be dangerous things.

Sadly, after the Egyptians, other mathematical cultures became a bit snobbish about solving linear equations, and refused to write down their methods for doing so. I assume they wanted to pretend to later generations that such problems were too easy for them to waste space on their clay tablets (or whatever) with detailed answers. I am sure that plenty of Babylonian, or Greek, or Chinese, or Indian schoolchildren disagreed.

---

55. The length of a fish is twice the length of its head, and then 10cm more. If the length of the fish is 22cm, what is the length of its head?

Things have changed since Egyptian times. New methods and techniques have grown up to avoid the need for a trial-and-error approach. The process of solving linear equations is now governed by a strict set of rules and regulations, which all have their roots in the logic of numbers, but which, when you're in the classroom, gazing through the window at the playing field outdoors, can appear as arbitrary as the laws of the cricket game that you wish you were playing.

Why do you have to do the same to both sides? Why do you have to multiply both numbers in the bracket? Why do you have to cross-multiply? It's not good enough that Mr Barton mutters, 'Because you do, class, you just do,' and walks quickly away as his stress levels rise to dangerous heights.

Solving an equation is like a puzzle where the aim is to

isolate the unknown on one side, without breaking any of the unwritten rules of numbers. All of the commands mentioned in the paragraph above are the names of possible moves in this perverse game.

Take this complicated linear equation.

$$\left(\frac{2(x + 3)}{x} \right) = \left(\frac{3}{4} \right)$$

Over time, people have figured out that the best thing to do to start off with in equations of this sort is to 'cross-multiply' in order to get rid of the fractions. 'Cross-multiplying' is just a special name for a common kind of 'doing the same thing to both sides'.

'Doing the same thing to both sides' is the most common tactic that we are taught to use in classrooms nowadays, and is based on the understanding that both sides of the equation are equal. So, if you add the same thing to both sides, or subtract the same thing from both sides, or multiply both sides by the same thing, or divide both sides by the same thing, you will always end up with another equation where both sides are still perfectly balanced. The problem with complicated equations like the one above is that each side is itself quite complicated. You have to follow the correct sequence of moves, otherwise you end up getting further and further out of your depth.

Cross-multiplying turns out to be the correct first move in this case. It really involves two stages. First, you 'do the same thing to both sides' by multiplying both sides of the equation by four. This move will get rid of the fraction which has a denominator of four:

$$4 \times \left(\frac{2(x + 3)}{x} \right) = \left(\frac{3}{4} \right) \times 4$$

Note that, when you are multiplying both sides by four, it is necessary to treat each side as an individual entity. You have to multiply the whole of $\dfrac{2(x+3)}{x}$ by four, and the whole of 3/4 by four. You can't just multiply a particular bit of one side by four unless you can prove that this is a valid move. For example, when multiplying the left-hand-side by four, you can't just decide, for the sake of making your life easier, that the multiplication only applies to the 3 inside the bracket and write: $\dfrac{2\,[x + (4 \times 3)]}{x}$

because then you will not have multiplied the WHOLE of the left-hand-side by four, and maintained equilibrium in the equation.

It is possible to simplify the left-hand-side, since

$$\frac{3}{4} \times 4 = \frac{3}{4} \times \frac{4}{1} = \frac{12}{4} = 3$$

So the equation is now: $4 \times \left(\dfrac{2(x+3)}{x} \right) = 3$

Next you 'do the same thing to both sides' by multiplying both sides of the equation by x. This is the kind of thing that makes people feel uneasy, because you don't know what number x is. However, you can't argue with the fact that, whatever number x is, as long as you multiply both sides of the equation by it, you will end up with another equation where both sides are still balanced. In addition to this, you will be able to get rid of the fraction which has a denominator of x:

$$x \times \left(4 \times \left(\frac{2(x+3)}{x} \right) \right) = 3 \times x.$$

At all times, you have to be careful that you do not make a move that breaks the rules of numbers, but since, for any three numbers, it does not matter what order you multiply them, or which two numbers you decide to multiply first (i.e. $3 \times (4 \times 5)$ $= (3 \times 4) \times 5 = 5 \times (4 \times 3) = (4 \times 3) \times 5$), the right-hand-side of the equation can be simplified:

$$x \times \left(4 \times \left(\frac{2(x + 3)}{x} \right) \right) = 4 \times \frac{x}{1} \times \frac{2(x + 3)}{x} = 4 \times \frac{2x(x + 3)}{x} = 4 \times 2(x + 3)$$

So the equation is now: $4 \times 2(x + 3) = 3 \times x$.

Cross-multiplying is a shortcut for equations of this type, where each side is a fraction, and carries out two moves in one go. You multiply the numerator of the left-hand-fraction by the denominator of the right-hand-fraction, and the numerator of the right-hand-fraction by the denominator of the left-hand-fraction. It is normally represented like this:

$$\frac{2(x + 3)}{x} = \frac{3}{4} = 4 \times 2(x + 3) = 3 \times x$$

Again, the equation can be simplified:

$4 \times 2(x + 3) = 4 \times 2 \times (x + 3) = 8 \times (x + 3) = 8(x + 3)$
and $3 \times x = 3x$.

Therefore: $8(x + 3) = 3x$.

As a result of experience with dealing with this kind of problem, it is generally agreed that the best thing to do next is to 'remove the brackets', which involves multiplying both things inside·the bracket by the number outside it. The reason for this is that:

$8(x + 3)$
$= 8 \times (x + 3)$
$= (x + 3) + (x + 3) + (x + 3) + (x + 3) + (x + 3) + (x + 3) + (x + 3) + (x + 3)$
$= x + x + x + x + x + x + x + x + 3 + 3 + 3 + 3 + 3 + 3 + 3 + 3$
$= 8x + 24$

The argument relies on basic truths about multiplication and addition, but it is necessary in order to make absolutely sure that you are not making any false moves. It is now certain that $8(x + 3) = (8 \times x) + (8 \times 3) = 8x + 24$, and the new equation reads: $8x + 24 = 3x$.

From here, it is a matter of several applications of 'doing the same thing to both sides' in order to isolate the x. First you subtract 24 from both sides: $(8x + 24) - 24 = 3x - 24$.

But $(8x + 24) - 24 = 8x$ (since the number that is 24 less than the number-that-is-24-more-than-$8x$ is $8x$), so: $8x = 3x - 24$.

Then you subtract $3x$ from both sides: $8x - 3x = (3x - 24) - 3x$.

Since there is no difference between subtracting 24 from $3x$, and then subtracting $3x$, or, subtracting $3x$ from $3x$ and then subtracting 24:

$$(3x - 24) - 3x = (3x - 3x) - 24 = 0 - 24 = -24.$$

Also $8x - 3x = 5x$.

And so the equation reads: $5x = -24$.

Finally you divide both sides by 5: $\dfrac{5x}{5} = \dfrac{-24}{5}$

But if you share $5x$ between 5 people, each person gets x and so $5x/5 = x$. Therefore:

$$x = \dfrac{-24}{5}$$

You have isolated x, and the equation is solved.

I have gone through the above example in far more detail than is normally supplied when such equations are solved in textbooks, or by mathematics teachers, and much of the argument above may appear self-evident. However, I want to show that, at each stage, the steps for solving linear equations rely on truths about the way numbers in general work. It might appear obvious that it does not matter in what order you multiply three numbers, but the same thing is not true when, for example, you subtract three numbers: $9 - (2 - 1)$ is not the same as $(9 - 2) - 1$ (where the brackets tell you which subtraction to do first).

> **56.** You notice that there is twice as much of the day left as has already gone (considering the day as being twenty-four hours long and to start at midnight). What time is it?

Solving linear equations in this way is like a maze. There is not just one way to get to your final destination, but there does tend to be a route which is simplest. In the above problem, by making the moves in the particular order that you have made them, you have generally avoided having to deal with anything particularly unpleasant. However, I know that people don't like to feel that they have no choice in life, and so an alternative path to the answer might run as follows:

$$\frac{2(x + 3)}{x} = \frac{3}{4}$$

$$8(x + 3) = 3x \qquad \text{(cross–multiplying as before)}$$

$$\frac{8(x + 3)}{8} = \frac{3x}{8} \qquad \text{(dividing both sides by 8)}$$

$$(x + 3) = \frac{3x}{8} \qquad \text{(simplifying)}$$

$$(x + 3) - 3 = \frac{3x}{8} - 3 \qquad \text{(subtracting 3 from both sides)}$$

$$x = \frac{3x}{8} - 3 \qquad \text{(simplifying)}$$

$$x - \frac{3x}{8} = \left(\frac{3x}{8} - 3\right) - \frac{3x}{8} \qquad \text{(subtracting } \frac{3x}{8} \text{ from both sides)}$$

$$\frac{5x}{8} = -3 \qquad \text{(simplifying)}$$

(This simplification is hard, and where this route to the solution becomes more complicated than the other one. You have to subtract three-eighths of x from one whole x. One whole x contains eight-eighths of x. If you subtract three-eighths of x from eight-eighths of x, you are left with five-eighths of x:
$x - \frac{3x}{8} = \frac{8x}{8} - \frac{3x}{8} = \frac{5x}{8}$.)

$$8 \times \frac{5x}{8} = -3 \times 8 \qquad \text{(multiplying both sides by 8)}$$

$$5x = -24 \qquad \text{(simplifying)}$$

$$\frac{5x}{5} = \frac{-24}{5} \qquad \text{(dividing both sides by 5)}$$

$$x = \frac{-24}{5} \qquad \text{(simplifying)}$$

You have reached the same solution as before, but in this case it was necessary to deal with complicated fractions involving unknowns, where previously you managed to avoid such problems.

It is very easy to make mistakes when solving linear equations, because the rules of generalised numbers are not easy to follow. For example, the following step seems to be a perfectly reasonable thing to do:

$$\frac{2x + 6}{x} = \frac{3}{4}$$

$$\left(\frac{2x + 6}{x}\right) - 6 = \frac{3}{4} - 6 \qquad \text{(subtracting 6 from both sides)}$$

$$\frac{2x}{x} = \frac{3}{4} - 6 \qquad \text{(simplifying)}$$

However, if you continue on to try and solve the problem now, you will find yourself with all sorts of problems. Something must have gone wrong somewhere.

There is nothing wrong with trying to subtract six from both sides, but there is an error in the simplification process. In the expression $\left(\frac{2x + 6}{x}\right) - 6$ you have simply removed the sixes,

arguing that $6 - 6$ is 0. However, this is not valid because the first six does not stand on its own – it is divided by x. In this case, you cannot be certain that the move you have made follows the general rules of number.

---

57. A fisherman is desperate to drum up business, and comes up with the following special offer. On days when he has caught a fish, he will sell it to a customer for £3, but on days when he has caught no fish, he will give the customer £2 just for the trouble of coming to his shop. Sadly, the fisherman is not very skilled, and over the coming month, he never catches more than one fish in a day. At the end of the month (a period of thirty days), he finds that he has paid out the same amount as he has received, which prompts him to hang up his galoshes and become an investment banker. How many days did he catch a fish?

3   CHANGE ALL THE SIGNS

One of the rules of number that always confuses people is 'subtracting a bracket'. The rule that governs this process says that if you are faced with an algebraic expression or equation in which it is necessary to subtract a bracket, you remove the brackets and change the signs of everything inside it. So, for example, if you have $a - (b + c)$, that is the same as $a - b - c$, and if you have $a - (b - c)$ that is the same as $a - b + c$. Is it possible that this fact is based on some rational and logical truth?

It is best first to think in terms of particulars. Take $10 - (2 + 3)$. But don't combine the 2 and the 3, treat them as separate entities. In words, you could translate this sum as: 'take three more than two away from ten.' You could do this in two stages, by first taking away two from ten, and then taking away another three from ten. You could even draw a picture that looked something like this:

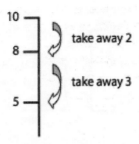

It seems that $10 - (2 + 3)$ is the same as $10 - 2 - 3$, which agrees with the general rule above.

However, you could translate the sum into exactly the same words, and draw exactly the same picture, if, instead of actual particular numbers, you had letters that represent general numbers. The point is that, no matter what numbers you are given, the overall effect is the same.

So, take the general sum $a - (b + c)$. You could translate this statement into words as: 'Take away c more than b from a'. And, again, you could draw a picture:

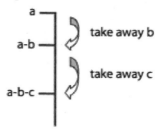

It appears that it is quite true that $a - (b + c) = a - b - c$. There is no difference between the process behind each sum. In fact, the first example is just a particular case of the second.

The second case is always more confusing. $a - (b - c) = a - b + c$. Take $10 - (5 - 3)$. This sum could be translated into words as: 'Take away three less than five'. In other words, if you go ahead and take away five, you have taken away too much. You wanted to take away three less than five. You rushed in like a bull in a china shop, and now you need to add three to get to where you were asked to go. Here is a picture:

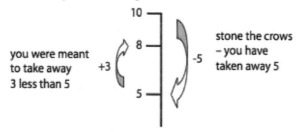

So it seems that $10 - (5 - 3)$ is in fact the same as $10 - 5 + 3$. And, just as before, whatever the numbers you use, the method of working out the sum will remain the same, and so you replace the numbers with generalised terms. $a - (b - c)$ can be interpreted as: 'Take c less than b from a'. In other words, if you take b away you have gone too far, and must add c to get to where you were meant to be:

you were meant
to take away
c less than b

$+c$

$a-b+c$

$a-b$

$-b$

strike a light
– you have
taken away b

It might be complicated and confusing, but at least algebra is based on something that resembles commonsense.

---

58. The length of a rectangle is twice its width. If the perimeter is 36cm, find the width.

---

4 FALSE ASSUMPTIONS

The Babylonians might have felt that an explanation of how they solved linear equations was beneath them, but they weren't as coy about talking about simultaneous equations. They had their own way of dealing with them, and their methods were no doubt quite different to what you learnt at school. I don't want to dredge up those memories, but perhaps you would like an alternative to the standard approaches.

A Babylonian problem might have gone something like this:

'One of two fields yields 4 sila per sar, and the second yields 3 sila per sar. [This is actually a lot of sila for a sar to yield, but I'm hoping you won't spot that, and it simplifies the problem a bit.] The yield of the first field is 190 sila more than the yield of the second field, and the area of the two fields combined is 100 sar. Find the area of each field?'

If you take this information, and put it in equation form, (which the Babylonians would not have done), you should get two equations. Call x the area of the first field in sar, and y the area of the second field in sar. Then, if the first field yields 4 sila for every sar, its overall yield will be $4 \times x$ (or just $4x$), and, if the second field yields 3 sila per sar, then its overall yield will be $3 \times y$ (or $3y$). The problem states that the yield of the first field is 190 more than the yield of the second field. In algebraic terms this can be translated as: $4x - 190 = 3y$, or $4x - 3y = 190$. The problem also states that the sum of the areas of the two fields is equal to 100, and therefore $x + y = 100$.

The Babylonians were just as casual in their methods as the

Egyptians. They also made a false assumption – something we have all done at one time or other: I once falsely assumed that my parents were asleep, and suffered a psychologically damaging episode as a result. Anyway, the Babylonians would probably have falsely assumed that both fields had an area of 50 sar, because this is a simple starting point, and is a solution to the second equation ($50 + 50 = 100$), although not the first. If both fields have an area of 50, then the value of $4x - 3y$ is 50 [(4×50) – (3×50)], which is 140 out from the required answer of 190.

But, the Babylonians were quick to remedy their false assumptions. They fiddled with their initial guess. They realised that, in order to keep the second equation happy, if they increased their guess for x by 1, they must decrease their guess for y by 1. For example, if they decided to try x as 51, then they must try y as 49, since $51 + 49 = 100$.

Here are some examples of the consequences of the Babylonians' fiddling.

x	y	$x + y$	$4x$	$3y$	$4x - 3y$
50	50	100	200	150	50
51	49	100	204	147	57
52	48	100	208	144	64
53	47	100	212	141	71

In each example, the sum of x and y is 100, which means that the second equation in the problem is always satisfied.

As x increases in steps of 1, y must correspondingly decrease in steps of 1.

As x increases in steps of 1, $4x$ increases in steps of 4, and as y decreases in steps of 1, $3y$ decreases in steps of 3.

Since $4x$ is increasing in steps of 4, and $3y$ is decreasing in steps of 3, and you are subtracting $3y$ from $4x$, then $4x - 3y$ must increase in steps of 7, because, as you move down the rows in the table, you are taking away a number that is 3 less from a number that is 4 more.

Now, for the Babylonians' initial guess, where x and y were both 50, the value of $4x - 3y$ was 50, which is 140 short of the required 190. They now know that every time they increase the value of x by 1, then the value of $4x - 3y$ increases by 7. It is a question of working out how many increases of 7 will make up the shortfall of 140, and the answer to this question is 20.

Therefore, if the initial guess for x is increased by 20 (so that $x = 70$), and the initial guess for y is correspondingly decreased by 20 (so that $y = 30$), they will have solved the problem. And that is exactly what they did. The first field has an area of 70 sar and the second field has an area of 30 sar.

---

59. There are a number of princes and frogs in a fairy-tale. In total they have thirty-five heads and ninety-four feet. Find the number of each.

5 THE LOGIC BEHIND SIMULTANEOUS EQUATIONS

I lied when I said that I would not put you through the pain of dealing with the methods for solving simultaneous equations that you learnt at school. But I hope that, in what follows, those methods will be revealed to be based on principles of sound commonsense – just like the methods that the Babylonians used.

Nowadays, fields and their yields are out of favour with the writers of textbooks. It is the prices of cups of coffee and tea that are in fashion. So your problem is to find the price of one cup of coffee and the price of one cup of tea, given that one cup of coffee and one cup of tea cost 100 pence, and four cups of coffee and two cups of tea cost 280 pence.

If you call x the price of a cup of coffee, and y the price of a cup of tea, these two pieces of information lead to two equations. The first statement is equivalent to the equation $x + y = 100$, and the second statement is equivalent to the equation $4x + 2y = 280$.

The most common method taught in schools is to solve two equations like this by elimination, and the first stage in this method is to play around with one of the equations so that the numbers in front of one of the variables are the same. (Numbers in front of variables are called coefficients.)

If you take the first statement ('the price of one cup of coffee and one cup of tea is 100 pence'), it is possible to derive further pieces of information about the prices of coffee and tea from it. For example, if one coffee and one tea cost 100 pence, then two coffees and two teas cost 200 pence, because you have simply doubled your order. Furthermore, three coffees and three teas cost 300 pence, and four coffees and four teas cost 400 pence.

Each of these statements is equivalent to an equation in which each of the parts of the original equation is multiplied by the same number. If you multiply $x + y = 100$ by 2 on both sides, you get $2x + 2y = 200$, if you multiply $x + y = 100$ by 3 on both sides, you get $3x + 3y = 300$, and if you multiply $x + y = 100$ by 4 on both sides, you get $4x + 4y = 400$.

Given that all these statements are equivalent to one another, you can replace the original statement with any one of them. Since the aim is to alter the first equation so that the coefficients in front of one of the variables are the same, and since the coefficient in front of the x in the second equation is 4, you replace the statement 'one cup of coffee and one cup of tea cost 100 pence' with the statement 'four cups of coffee and four cups of tea cost 400 pence'.

Algebraically, this is the same as replacing the equation $x + y = 100$ with the equation $4x + 4y = 400$. You are then in a situation where both equations ($4x + 4y = 100$ and $4x + 2y = 280$) have the same coefficient in front of the x. Back in your exercise books, this step will have looked like this:

Equation 1	$x + y = 100$
Equation 2	$4x + 2y = 280$
4 × Equation 1	$4x + 4y = 400$
Equation 2	$4x + 2y = 280$

So far, so good. You are using two pieces of information: 'fours cups of coffee and four cups of tea cost 400 pence' and 'four cups of coffee and two cups of tea cost 280 pence'. The first statement deals with an order of four cups of coffee and four cups of tea, and the second statement deals with an order of four cups of coffee and just two cups of tea. Two cups of tea is the difference, and, since the difference in price of the two orders is 120 pence, two teas must cost 120 pence.

This logical step is the same as subtracting the second equation

$(4x + 2y = 280)$ from the first one $(4x + 4y = 400)$ to get the simpler equation $2y = 120$:

$$\begin{array}{r} 4x + 4y = 400 \\ - \ \underline{4x + 2y = 280} \\ 2y = 120 \end{array}$$

This is the algebraic equivalent to finding the difference in drinks ordered between the two statements, and equating it to the difference in price. Because you have the same number of cups of coffee (i.e. the same number of 'x's), you end up with a statement and related equation that only deal with the price of tea.

Once you have reached this stage, your problems are almost over. If two teas cost 120 pence, then one tea must cost 60 pence. Now you can go back to the original statements to work out the price of one cup of coffee. If one tea costs 60 pence, and one tea and one coffee cost 100 pence, then one cup of coffee must cost 40 pence. You can then check these answers also work for the second statement. If tea costs 60 pence and coffees cost 40 pence, then four coffees and two teas cost 280 pence. Problem solved.

Once again, you have relied solely on logic to guide you to the answer, but at school you framed the argument above in its algebraic form. It will have looked something like this:

$$2y = 120$$
$$y = 60$$

Substituting in Equation 1: $\qquad x + 60 = 100$
$$x = 40$$

Check in Equation 2: $\qquad 4x + 2y = 280$
$$(4 \times 40) + (2 \times 60) = 280$$
$$280 = 280 \quad \text{(TICK)}$$

60. As a student, you used to supplement your income by visiting the sperm bank. Forty years later, it turns out that, in addition to the son you have through your marriage, you have another biological son as a result of these visits, and so you decide to change your will. You have £10 000 to share between them, and you want a fifth of the inheritance of the son you always knew about to be £1100 more than a quarter of the inheritance of the son you have only just found out about. How much does each son get?

It is perfectly possible to replace the first equation with an equation so that the coefficients of y are the same. In this case, you replace 'one coffee and one tea cost 100 pence' with 'two coffees and two teas cost 200 pence'.

The two statements you are now dealing with are 'two coffees and two teas cost 200 pence' and 'four coffees and two teas cost 280 pence'. In this case the difference in the orders is two coffees, and the difference in price is 80 pence. Therefore, two coffees cost 80 pence, and one coffee costs 40 pence. If one coffee costs 40 pence, and one coffee and one tea cost 100 pence, then one tea must cost 60 pence. If one coffee costs 40 pence and one tea costs 60 pence, then 4 coffees and two teas cost 280 pence. You have checked that the second statement is also true for these prices, and once again the problem is solved.

Algebraically the argument above would be represented as:

Equation 1	$x + y = 100$
Equation 2	$4x + 2y = 280$
$2 \times$ Equation 1	$2x + 2y = 200$
Equation 2	$4x + 2y = 280$

$$\text{Equation } 2 - (2 \times \text{Equation } 1)$$

$$\begin{aligned} 4x + 2y &= 280 \\ - \quad 2x + 2y &= 200 \\ \hline 2x &= 80 \\ x &= 40 \end{aligned}$$

Substitute x in Equation 1: $40 + y = 100$

$y = 60$

Check in Equation 2: $4x + 2y = 280$

$(4 \times 40) + (2 \times 60) = 280$

$280 = 280$ (TICK)

I am sorry to have exhumed such problems from long-forgotten graveyards in your memory, but I hope that you can now see that this method for solving simultaneous equations relies on a series of logical steps. Each step is mirrored by an algebraic procedure. Once these procedures are accepted as valid, then the use of algebra speeds up the solution of the problem, but the procedures should never be allowed to obscure the rational basis for what you are doing.

6 SQUABBLING SCHOOLBOYS

Once you have got a taste for algebra, it is hard to stop. The ancient Egyptians, as we have seen, were perfectly happy to take on linear equations in one variable, since there are plenty of examples of problems of this type in the Rhind Papyrus. At approximately the same time (1900-1650 BC) the Babylonians had gone on to tackle systems of two simultaneous linear equations involving two variables, using their false assumptions.

The Babylonians didn't stop there, but went on to solve certain types of quadratic equations in one variable. Such equations contain only one unknown, and no powers of this unknown greater than 2 – so you can have terms involving x^2, but not terms involving x^3 or higher powers of x (e.g. $2x^2 + 3x - 4 = 0$). In some instances, the Babylonians even dared approach equations with higher powers of x. Their enthusiasm was apparently boundless. The Chinese, the Indians and the Arabs also made important discoveries in the field of algebra, and reached levels of sophistication that Europeans did not achieve until many centuries later.

During the 13th century, Chinese mathematicians like Chin Chiu Shao were happily solving highly complicated algebraic problems. Chin was a sage from the Shin Jour province who in 1247 composed a book called *The Nine Sections of Mathematics* which happily gave advice on problems involving calendar calculations, the weather, field-surveying, taxation, fortification works, construction, military affairs and business, as well as a scheme for solving numerical equations. He also found time to gain a reputation for being an insatiable lover and an excellent polo-player. Europeans, in the meantime, were mostly fighting each other in a series of bloody wars, or living in abject poverty.

However, Europe could not hide from 'x's and 'y's forever. Algebra is apparently contagious and Italy began to contract it towards the end of the 13th century, when the mathematical works of the Arabs began to find an audience amongst the new merchant classes. The most influential of these works were translations of al-Khwarizmi's introduction to algebra: '*Al-kitab al-muhtasar fi hisab al-Jabr wa-l-Muqabala*'. The title translates as 'The Condensed Book of the Calculation of al-Jabr and al-Muqabala'. The term '*al-Jabr*' has already been explained. 'Al-Muqabala' is the process of simplifying an equation by removing the same quantity from each side. For example, it is possible to simplify the equation $5x + 2 = 4$ by taking away two from both sides, so that it becomes the equation $5x = 2$.

Once they had got the bug, European mathematicians became huge fans of solving equations. Al-Khwarizmi and his friends had already showed how to solve any kind of quadratic equation, so the Europeans had a crack at solving any cubic equation in one unknown (an equation where the highest power of x is x^3). A team of Italian mathematicians, called Scipione del Ferro (1465-1526), Niccolò Tartaglia (1500-1557) and Gerolamo Cardano (1501-1576) had achieved this by the end of the 16th century, and it did not take long for a pupil of Cardano, called Ludovico Ferrari (1522-1565), to show how to solve any quartic equation in one unknown (an equation where the highest power of x is x^4).

It did not stop there. Like overexcited schoolboys, later scholars tried to find ways of solving any quintic equation (yes, that is an equation where the highest power of x is x^5). Sadly for them, it was eventually shown early in the 19th century, by a Norwegian called Niels Abel (1802-1829) and a Frenchman called Evariste Galois (1811-1832), that it was not possible to come up with a general formula for solving such equations.

61. A group of friends decide to club together to buy a hammer-action electrical drill as a wedding present. They find that if they each pay $8, they will have paid $3 too much, and if they each pay $7, they will have paid $4 too little. How many friends were there?

The similarity to schoolboys does not end there. Mathematicians of the time were very competitive with one another, and were constantly challenging each other to contests to find out which one of then was top of the class. Scipione del Ferro was the Chair of Arithmetic and Geometry at the University of Bologna, but his position of authority did not encourage him to behave in a responsible manner. He kept his methods for solving cubic equations to himself so that he could win any competitions that came his way, as the prizes for these were often considerable sums of money or posts at prestigious universities. He did, however, confide his techniques to one of his students, Antonio Maria Fiore. At the same time, one of del Ferro's contemporaries, Niccolò Tartaglia, was going around boasting that he, too, could solve cubic equations. If you did that in a school playground nowadays, you would be in the nurse's office with two black eyes in seconds.

Eventually, there was a showdown. In around 1535, Tartaglia and Fiore challenged each other to a mathematics competition. Each man gave the other thirty problems to solve. In a high-risk strategy, Fiore's problems all reduced to solving the same kind of cubic equation – the kind which del Ferro had told him how to deal with. Tartaglia's problems for Fiore were more of a mixed bag. Unfortunately for Fiore, Tartaglia was able to solve the type of cubic equation that he had given him, which meant that he scored rather high marks in his test. Fiore, on the other hand was less fortunate, and Tartaglia was declared the winner, magnanimously declining the prize, which was that the loser should prepare thirty banquets for the winner and his friends. I suspect that Tartaglia was feeling fairly smug.

7 ALGEBRA IS DEMOCRACY

Even though in the 16th century mathematicians like Fiore and Tartaglia were gradually pushing back the boundaries of equations, they had still not developed algebra as we understand it today.

As we have seen, algebra is essentially the ability to generalise, and the process of generalising mathematics had started way back with the Pharaohs, when Egyptian mathematicians started to talk about an unknown number as a 'heap'. Plenty of other cultures took this step. The Babylonians called an unknown number, 'ush' (which means 'length'), and they had other words for things like the square of an unknown number, which they called 'sagab' (which does in fact mean 'square'). The Indian mathematician Brahmagupta (598-670 AD) used the abbreviated words for colours to denote unknown numbers.

The next step was to apply symbolisation more widely. Again, this happened in various different places at various different times. The Greek mathematician Diophantus, who lived from around 200 BC to 284 BC, developed a short-hand for expressing mathematical terms in approximately 250 AD. He had symbols to represent unknowns and numbers, but the terminology was still quite complicated. He would have written $\Delta\gamma\gamma\varsigma\iota\beta M\theta$, which translates as x^2 3 x 12 units 9, or $3x^2 + 12x + 9$.

Because such terminology was so complicated, it meant that it was difficult to see underlying patterns in the way people solved mathematical problems, and it was impossible to express such patterns in a simple way. In general, mathematicians would simply write down many examples of how to solve a particular type of problem, and hope that their readers would understand the theory behind it. In 1629, the famous mathematician and philosopher René Descartes complained: 'Algebra, if only we could extricate it from the vast array of numbers and inexplicable figures by which it is overwhelmed, so that it might display the clearness and simplicity which we imagine ought to exist in a genuine Mathematics.'★

---

62. A man has two perfumes, one of which sells at ten pounds a bottle, and the other of which sells at four pounds a bottle. What is the mixture that would sell at six pounds a bottle?

It was a man called François Viète from Poitou (1540-1603) who finally developed a sufficiently clear mathematical language. Having spent his life falling in and out of favour with the French kings of the second half of the 16th century, at a time when France was in turmoil as religious groups fought against each other, Viète eventually gained the support of Henry IV, who used his talents to try to decode the messages that were being sent to his enemies by Philip II of Spain. Viète was successful – so much so that Philip complained to the Pope that Black Magic was being used against him.

During those periods of time when he was not in favour, Viète devoted himself to his study of mathematics. In terms of algebra, his major breakthrough was to generalise not just unknowns in an equation, but also to develop symbols for operations like addition (he introduced + for addition and − for subtraction), and to generalise the numbers in front of the unknowns. In other words, where previous scholars might deal with a whole series of similar equations separately (e.g. $2x + 3 = 10$, $4x - 1 = 13$, $6x + 23 = 132$ and so on), Viète could talk about all of these individual equations as being of a general type $ax + b = c$, where a, b and c stand for the numbers in the particular examples. He could then explain, in general, how to

---

★ Descartes (1997) p15.

solve all types of the same equation much more concisely than his predecessors.

This might not seem overexciting, but it was highly significant at the time as it enabled mathematicians to write down generalised formulas to solve particular types of problems. Up to this point, the Arabs and others had been able to solve any type of quadratic equation that you might fancy giving them, but they had done this by classifying quadratic equations into five different types, and coming up with different methods for solving each. For example, they had one method for solving equations of the general form $bx = ax^2$ (e.g. $2x = 3x^2$, $4x = 6x^2$, or $22x = 9x^2$), but a completely different way of handling equations of the general form $ax^2 + bx = c$ (e.g. $4x^2 + 2x = 1$, $5x^2 + 4x = 7$ or $6x^2 + 2x = 11$). The Arabic techniques were very ingenious, but due to their connection of algebra with geometry and apathy towards negative numbers, they failed to see that all quadratic equations could be solved in the same way.

Once Viète had developed his form of algebra, however, it was possible to see that all quadratic equations could be considered to be of one type: $y = ax^2 + bx + c$ (where a, b and c are the numbers in front of the x^2 term, the x term, and at the end of the equation). From here, a general formula could be given for solving any quadratic equation. Just to give you the pleasure of meeting an old friend, that formula is:

$$y = \frac{-b \pm \sqrt{b^2 - 4ac}}{2a.}$$

Again, I will forgive you for being underwhelmed, but Viète's discovery had profound effects on the lives of his contemporaries and everyone ever since. Previously, mathematics had been the preserve of an educated elite made up of scholars or priests, and had been viewed with suspicion by the average man, mostly because it was simply too complicated for him to even begin to follow. However, with the growth of commerce in Europe during the 17th century, the practical value of more advanced mathematics became apparent. Algebra provided craftsmen and businessmen a way of accessing mathematics. They were able to use formulae to solve practical problems, even if they did not fully understand where they came from. Mathematics was beginning to be seen as a tool for the people, rather than a mysterious black magic practiced by the elite.

In addition to this process of demystification, algebra allowed complex situations to be analysed and manipulated. It would not be long before Galileo Galilei (1564–1642) started to investigate laws that govern motion, or Johannes Kepler (1571–1630) studied the movements of the planets, or Isaac Newton (1643–1727) watched apples falling from trees. In our own century, we have managed to navigate to the moon and to other planets, and Albert Einstein (1879–1955) has come up with his theory of relativity. Without the simplifying power of algebra, and its ability to highlight relationships and patterns, the work of these men would have been impossible.

8 THE SAVING OF CHARLIE

I am not sure that back in the classroom Charlie is convinced. He is still in all kinds of trouble. He just can't get his head around the concept of a t, or an x, or a y, applying to anything of significance, and faced with more than one of them in any given expression, he sinks into a pit of self-loathing and despair. So, maybe it would be helpful for Charlie to have a look at how people were coping with algebra before mathematicians like Viète came along and purified it into such a puzzling form…

No one should feel ashamed that they require a bit more substance to their maths problems. Until the time of Viète, most mathematicians had trouble understanding equations unless they linked them to geometrical shapes. That is why the Babylonian word for an unknown number was 'line', and why we talk of 'x squared' and 'x cubed'. Viète himself still referred to his unknowns in terms of cubes, planes and lines. It is just much easier to deal with an equation if you can draw a picture of it. So here is how the Franciscan monk Luca Pacioli (or Brother Luca), approached equations using diagrams.

Pacioli is often revered by accountants all over the world as the 'Father of Accounting', because he gave the first full description of double-entry bookkeeping as part of a work entitled *Everything about Arithmetic, Geometry and Proportions*, which was published in 1494. He liked to say that 'a person should not go to sleep at night until the debits equalled the credits'!*

---

* Geijsbeek (1914).

However, for now, we must restrain ourselves from being sidetracked into a study of the finer points of accountancy. We are concerned to see how Pacioli went about solving a problem like $x^2 + 10x = 39$. He didn't wheel out massive formulae, or get stuck into complicated procedures. He came up with a way of considering the problem in terms of lines and squares.

First he drew this picture:

Square A has an area of 'x times x' (or just x^2). Rectangles B and C both have an area of '5 times x' (or just $5x$), and so their combined area is $5x + 5x$ (or just $10x$). Therefore, the combined area of A, B and C is $x^2 + 10x$. Pacioli has drawn a geometrical picture which is equivalent to the equation that he is trying to solve.

Once he has done this, then solving the equation becomes a shape problem rather than an algebraic problem. In order to find the value of x, it is necessary to complete the picture by drawing in square D:

In this diagram, the combined area of A, B and C is $x^2 + 10x$, which is equal to 39 (according to the initial equation).

Square D has an area of '5 times 5', which is 25.

Therefore, the combined area of A, B, C, and D is 39 + 25, which is 64.

However, it is also true that the combined area of A, B, C, and D is a square, and so, in order to have an area of 64, it must have a side of length 8.

But, according to the diagram, the side of this square is also (x + 5), and therefore x must be equal to 3.

People like Pacioli spent a lot of time trying to change equations into shape problems with varying success. There are problems with such an approach. Firstly, each type of equation needs a different technique to solve it. Pacioli could solve any equation of a form similar to the one above (i.e. $x^2 + a\mathrm{x} = b$, where a and b are positive whole numbers) using the same kind of picture, but he had to find a totally different approach in order to solve a quadratic equation of a different form. In contrast to this, as you have seen, later mathematicians, who solved equations using 'pure' algebra, could make use of the dreaded 'formula' to solve any quadratic equation. Secondly, Pacioli's method fails to find negative solutions to equations, because, if you are dealing with shapes, a negative length has no meaning. In fact, the equation above has two solutions. One of them is 3, and the other is –13. Pure algebraic techniques will find both of these.

---

63. Just for kicks, this is the diagram which the Arab mathematician al-Khwarizmi used to deal with exactly the same equation as we looked at above: $x^2 + 10x = 39$.

A is a square with sides of length x, and B and C are both rectangles with sides of length x and 2½. I will leave it to you to work out how to solve the equation from this information.

It is best not to go too much further down the algebraic road. Things start to get a bit strange. The path is less clear. There are booby traps and barbed wire, and 'Keep Out' signs. And beyond them? Beyond them lie the Quadratic Reciprocity Theorem, and cyclotomic integers, and transcendental numbers, and quaternions, and Noetherian rings. You are not ready.

PART FOUR

CHANCE WOULD BE
A FINE THING

1 HIGH EXPECTATIONS FOR PROBABILITY

It probably comes as a surprise to learn that there was a time when big things were expected of probability. Sure it started with people figuring out how to predict how many tails would turn up if you flip a coin a hundred times, but when these predictions turned out to be correct, people got carried away.

The early pioneers of probability hoped that it would allow them to glimpse the future. If probability could tell you roughly how many sixes you will roll in an evening's gambling, then perhaps it could tell you roughly what was going to happen to your country over the next hundred years. It was simply a matter of applying probability to human affairs. Hence the observation from mathematician Pierre-Simon Laplace (1749-1827) in 1814 that, 'The fall of empires which aspired to universal dominion could be predicted with a very high probability by one versed in the calculus of chances.' (From *Théorie Analytique des Probabilités* 1814.) Although, to be fair, he was actually talking about the collapse of Napoleon's empire, and that had already happened. So, no prizes to probability for that prediction.

Other people saw probability as a way of finally sorting out what was the right thing to do in life. If you could just put a number to how good a thing was, and then put another number to how likely what you were about to do was to achieve that thing, then you could multiply the two numbers together, and that would give you another number, which would be the expected goodness of doing your action. And then you could compare it with all the other things that you might do at that particular time, and the amount of good that they might be expected to achieve, and whichever course of action had the largest number attached to it, that would be what you should do.

Gottfried Wilhelm Leibniz, the German philosopher, mathematician and logician (1646–1716) put it another way. He is probably most well known for having invented the differential and integral calculus (independently of Sir Isaac Newton), but he also spent plenty of time thinking about important philosophical issues. In particular, he wanted to reach a point where all decisions could be made with reference to purely logical principles, and therefore avoid all those messy arguments about 'the right thing to do', which take up so much of our time.

In the course of his thinking, he helpfully came to the conclusion that, to work out which course of action would maximise the good, you must act in order to create the largest rectangle. This would make more sense if I were to tell you that the length of one side of the rectangle represents the amount of goodness aimed at by a particular course of action, and the length of the other side of the rectangle represents how likely it is that the course of action will actually achieve this aim. Then the size of the rectangle is given by the (amount of goodness aimed at) × (the likelihood of your action achieving the goodness). Your aim is to make the rectangle as big as possible.

make this space as
big as you can

likelihood of achieving
the goodness

amount of goodness aimed at

He had such good intentions. Philosophers generally do. But here is what happens when you put that theory into practice. Let's say that you are at a friend's house, and that you decide to stay the night. We won't go into how you made that decision, but it does have far-reaching repercussions, because, since it is a spur-of-the-moment kind of thing, you have not brought your washbag. You find yourself in the bathroom, and there, standing provocatively in front of you in a specially designed holder, are your friend's toothbrush and toothpaste.

Let us also add the following pertinent facts. One: you know that your friend is the kind of person who will have trouble dealing with the image of the bristles on his toothbrush massaging someone else's gums. Two: the frozen lasagne you had for dinner has left a very unpleasant aftertaste in the back of your mouth. What do you do?

Well, according to Leibniz you need to try and make the largest rectangle. So, first you have to think about the goods involved in what you are about to do. To make things easy, let us assume that there are only two choices facing you at the moment: to use the toothbrush or not to use the toothbrush. You could, of course, just ignore the toothbrush, and go out into the street and help an old lady across the road. But let's just focus on the task at hand.

If you use the toothbrush, presumably the good you are aiming at is dental hygiene. (I can't think of any other reason you would scrub your teeth.) And if you do not use the toothbrush, then you are motivated towards doing the right thing by your friends. The next step, after working out what goods are on offer as potential results of your action, is to work out just how good these two possibilities are. And here I find that I run into a bit of trouble. How do you put a number to the goodness of dental hygiene or the goodness of acting thoughtfully towards your friends?

Take a scale of 0 to 10, where 10 is a really fantastic good, (e.g. single-handedly saving the world from alien invasion), and 0 is not a good at all (e.g. setting your parents' house on fire and cackling merrily as it burns to the ground). Where does the good of dental hygiene fit in this scheme of things? How do you work out its value? And even if you don't try and take Leibniz's ideas quite so literally, is it possible to even say which is the better good to aim at?

Leaving that aside for the moment, the next stage is to work out the probability of your actions achieving the good they are aiming at. If you use the toothbrush, what are the chances that this will lead to the good of dental hygiene? You could probably get a rough figure for that from dentists, I suppose, but since they are always very keen to persuade you to see them as often as possible, I am not sure that I particularly trust their statistics.

And if you don't use the toothbrush, will that lead to the good of doing right by your friends? I suppose it must do, but presumably part of the point of aiming at this good is that it will hopefully ensure that your friend will behave towards you with similar thoughtfulness in the future. Perhaps he will refrain from using your towel to cleanse himself after a thoroughly sweaty session in the gym. However, your friend is never going to know that you decided against using his toothbrush, unless you tell him about the moral tussle that you went through in order to reach this decision. And then he will either be annoyed that you even considered using his bathroom utensils, or he will think you are weird for telling him such an uninteresting piece of information.

All of which leaves you alone in the bathroom, desperately visualising rectangles in the mirror, with a growing feeling that Leibniz really hasn't done you any favours at all. Whilst he is probably right in saying that considerations both about the goodness of your aim and about the chances of achieving that aim do come into decision-making, he doesn't actually make the process of choosing what to do any easier. It is certainly almost impossible to connect numbers to different courses of action. So I suggest you lock the door of the bathroom, turn on the taps to disguise the telltale sound of bristle on enamel, and scrub away merrily. No one is ever going to know.

---

64. You have five pairs of earrings, but they have all got mixed up in their box. How many earrings do you have to take out, before you can be certain that you will have a pair?

So, probability does not quite deliver on the lofty aims of its early supporters. It can't tell you what to do, and it can't give detailed information about the future. That did not stop people continuing to make impressive-sounding noises about it. Descartes, for example, claimed that 'it is a very certain truth that, when it is not in our power to determine what is true, we ought to follow what is most probable.'[*] For Cicero (106 BC - 43 BC), 'Probability is the very guide of life.' The 19th century American mathematician, Charles Sanders Peirce (1839-1914), stated that: 'All human affairs rest upon probabilities, and the same thing is true everywhere. If man was immortal he could be perfectly sure of seeing the day when everything in which he had trusted should betray his trust, and, in short, of coming eventually to hopeless misery. He would break down, at last, as every good fortune, as every dynasty, as every civilisation does. In place of this we have death.'[**] Which is meant to be a good thing, I think.

---

65. In a board-game for two players, each player takes turns to move his counter. The aim is to move 100 spaces to the finishing-line, and the first player to cross it is the winner. Each player obeys a different set of rules. Bert moves his counter forward three spaces every go, whilst Ernie rolls a dice, and moves his counter the number of spaces that it shows. Is Bert or Ernie more likely to win?

---

[*] Descartes (2000) p57.
[**] Peirce (1955) p162.

2 IT'S A LOAD OF BALLS

The problem with applying probability to real-life problems is that real life is just so complicated. At any one time, we are faced with all sorts of choices, and it is often impossible to tell which is the best thing to do. Life wouldn't be very interesting if decisions were that easy.

Games, however, are a different kettle of fish. A game is usually controlled by very strict rules which are necessarily fairly simple. In addition, people want to win games, and so they are prepared to put a lot of thought into how they work in order to gain an advantage over their opponent. This is where the study of probability started.

The first man to begin to analyse games was the Italian mathematician Gerolamo Cardano. Cardano was the illegitimate child of Fazio Cardano, a man whose expertise in mathematics was such that he was consulted by Leonardo da Vinci on questions of geometry.

Initially taught by his father, Cardano studied medicine at university, boosting his finances by playing card games, dice and chess. In the 1530s Cardano began a period of intense mathematical investigation. During this time, in addition to his major contributions to algebra, Cardano made the first foray into the unexplored realms of probability theory. His book *Liber de Ludo Aleae* (*Book on Games of Chance* – which was written in the 1500s but not published until the 1660s) was the first study of things such as dice rolling, and was based on the premise that it is possible to apply scientific principles to games of chance. Some of the advice he provided on how to succeed in different games relied on concepts of probability – he examined the distribution of outcomes in, for instance, the sum of two dice thrown randomly – but some of it was purely practical. For example, Cardano points out that the chances of obtaining a particular card from a pack are greatly increased if you have previously rubbed it with soap.

The birth of probability may be traced back to the 16th century and Cardano was ahead of his time. But, ultimately, the questions that people like him began to answer using probability were the same kinds of questions that we were faced with in the classroom. And there is something comforting in the knowledge that men of mathematical genius have also pondered what the chances are of drawing a black ball out of a bag containing three black balls and two white balls. Or whether an odd or an even number is more likely to end up being thrown on a dice. Or what is the probability that Peter will pick a red sweet from a bag containing twenty-four green ones, twelve blue ones, and fifteen red ones. Well, maybe they didn't spend much time considering sweet problems, but they did spend a lot of time wondering about the other ones...

---

66. In a raffle at the end of the Carol Concert at the British Embassy in Kigali, there were 120 tickets in the hat. The prizes were a meal in the Intercontinental, a bottle of whiskey, and a plastic Christmas tree. What are the chances of winning a prize, if you bought just one ticket?

Pierre-Simon Laplace was another of the early students of probability. Born into a fairly prosperous farming family with little record of academic achievement in Normandy in 1749, Laplace gave up his degree and moved to Paris aged just nineteen to concentrate on his chosen subject, mathematics – and went onto to become the self-proclaimed 'best mathematician in France'.

Much of Laplace's mathematical career was spent trying to explain the irregularities in the motions of the planets. By this time, people had worked out that the planets moved in elliptical orbits around the sun, and they had come up with a mathematical model that could predict their movements. However, they were a little concerned that the planets sometimes behaved in slightly unexpected ways, although they always appeared to return to the predicted path. The thing that really worried them, was the possibility that these deviations might be becoming more and more pronounced. If this were the case, then perhaps there would come a time when the movements of the planets became so irregular that the whole fabric of our solar system would disintegrate, and the Earth would spin off into the darkness of deep space, with unpleasant side-effects for humans.

Laplace, however, analysed these irregularities, and showed that they were not becoming more serious, but that they formed part of a pattern. He essentially came up with a more complicated mathematical model that predicted the movements of the planets with greater accuracy, and calmed a lot of troubled minds in the process. However, he didn't entirely save the day, because one of his assumptions was that the planets are solid objects, when in fact their shapes change as they hurtle through space. This change in shape might lead to a gradual departure from the pattern of movement that we expect, but fingers crossed, everything will be all right for a while yet.

---

67. After the concert there was great excitement because the Ambassador produced a huge plate of twenty-five mince pies, sixty-four Ferrero Rocher, and forty-nine cocktail sausages. I was part of the general stampede towards the buffet, but I could not see the luxuries on offer, because the UN attaché for human rights had me in a headlock. Determined to take what was rightfully mine, I reached out blindly, and grabbed hold of something from the plate. Given that the mince pies, Ferrero Rocher, and cocktail sausages are mixed up at random, what is the chance that I don't get my hands on a sausage?

Anyway, Laplace took a bit of time off from his work on the planets to put down his thoughts on probability. 'The theory of chance consists in reducing all the events of the same kind to a certain number of cases equally possible, and in determining the number of cases favourable to the event whose probability is sought... The ratio of this number to that of all the cases possible is the measure of the probability, which is thus simply a fraction whose numerator is the number of favourable cases and whose denominator is the number of all cases possible.' (*Théorie Analytique des Probabilités*, 1814)

This is exactly the sort of problem you were given in the mathematics classroom. What is the chance of throwing higher than a 2 on a die? There are 6 equally likely possibilities, of which four are 'favourable' to the event of throwing higher than a two, and so the required probability is 4/6. What is the chance of a spinner landing on an odd number on a board with five equal sections numbered one to five? There are five equally likely possibilities, of which three are 'favourable' to the event of the spinner landing on an odd number, and so the required probability is 3/5.

Laplace himself demonstrates what he means by considering 'a large and very thin coin whose two large opposite faces, which we call heads and tails, are perfectly similar.' He uses his theory of probability to find the chances of getting at least one head after throwing this coin into the air twice. You might like to try and figure this one out for yourself, but don't be disheartened if you get stuck. This was cutting edge stuff at the time, and many mathematicians found the problem difficult.

3 MUDDY WATERS

Jean d'Alembert (1717-1783) was the illegitimate son of an artillery officer who – despite the distractions of rivalry with colleagues at the Paris Academy of Science and a general tendency to quarrel with all around him (he was known as 'a lightning rod which drew sparks from all the foes of the philosophers'[*]) made some outstanding contributions to the field of mathematics. For example, mathematical physicists had lost many hours of sleep in a controversy over the conservation of kinetic energy (the energy involved in motion), until d'Alembert stepped in, and improved Newton's definition of a force.

And yet, despite all this talent, he had all sorts of problems, when it came to solving Laplace's coin problem. He worked at it, and worked at it, and still he came up with the wrong answer, claiming that the probability of getting at least one head was 2/3. His argument relied on the fact that there were three possible combinations – a head and a head, a head and a tail, and a tail and a tail.

Laplace, very kindly and very gently, explains why d'Alembert was wrong. It all centres on the fact that the different 'cases' must be equally likely. In the examples in the previous chapter, a dice has equal faces, the sections on a spinner are equal sizes, and the coin has 'perfectly similar' faces. This means that all the numbers on the dice are equally likely, that the spinner is just as likely to land on any one number as any other, and that the

---

[*] Hawkind (1990) p3.

chances of the coin landing on a head are exactly the same as it landing on a tail. But there are plenty of situations where the different possibilities are not equally likely. When two football teams line up to face each other, each team knows that at the end of the match they will have either won, drawn, or lost, and yet these outcomes are not equally likely. That is why you can put money on the fact that San Marino will never appear at a World Cup. In the same way, you can buy dice that are loaded to land more often on a six. The probability of getting a six on such dice is no longer one in six.

However, that is the mistake d'Alembert made. It is true that, in a sense, there are only three possible combinations when flipping a coin twice. But if you think about it, you will see that one of them (a head and a tail) is more likely than the other two. The truth of this can be seen if you consider the possibilities for what can happen if you flip the coin twice. You can get a head then a head, a head then a tail, a tail then a head, and a tail then a tail. When d'Alembert talks about a head and a tail, he is actually combining two different possibilities: a head then a tail, and a tail then a head. He has not done what Laplace tells him to do, and that is to make sure that in a problem of this type, he is only considering cases that are 'equally possible'.

D'Alembert's mistake is an easy one to make, and there are other ones out there. For example, it is not true that, just because there is a one in six chance of rolling a three on a fair die, you will definitely roll a three, if you roll a die six times. It is only true that you will be unlucky if you do not. Probability does not tell you what will definitely happen, only what is likely to happen. Similarly, if you start rolling a die, and you get a three on your very first throw, that does not mean it is less likely that you will roll a three on your next throw. It makes no difference to the die what has happened in the past, it will not change its behaviour in the future. It, like us, does not learn from its mistakes.

> 68. Racy Rodders (RR) is twice as likely to win a horse race than High Hoof (HH) and High Hoof is twice as likely to win as Long Legs (LL). If only the 3 horses are in the race, what is the probability that Racey Rodders will not win?

I hope this has all made sense so far, because it is easy to get confused by probabilities. We are fairly good at assessing probable risks and what to do about them. For example, you probably avoid playing Russian roulette on a regular basis, for the sensible reason that the dangers involved are too great. However, it is much harder to work out how to take account of less likely threats. Which ones should you let affect the way you go about life, and which ones should you ignore?

In general, most people tend to feel that being struck by lightning is the kind of thing that it is best not to worry about. Records compiled in the US in 2002 by the National Safety Council show that the chances of death as a result of a lightning strike over your whole life are 1 in 56 439 (which is roughly the same as the chances of you being legally executed: 1 in 55 597). So, if you take this as a benchmark, here are some things that maybe you should worry a bit about, (although bear in mind that these statistics may differ from country to country): travelling as a pedestrian (odds of fatal injury 1 in 612), travelling as cyclist (odds of fatal injury 1 in 4587), travelling as a motorcyclist (1 in 1159), travelling in a car (1 in 228), climbing ladders or scaffolding (1 in 9175), drinking alcohol (1 in 10 493), and overexertion (1 in 29 101). And here are some things that perhaps you should worry less about: travelling in a bus (1 in 86 628), travelling by train (1 in 133 035), travelling in a three-wheeled motor vehicle (1 in 177 380), fireworks (1 in 744 997), dogs (1 in 206 944), snakes and lizards (1 in 1 241 661). I suggest you get in the market for a three-wheeled vehicle immediately.

With all these figures flying around, it is hard enough dealing with risk on your own, and so it doesn't help when other folk start adding to your problems. You have to be extremely careful when faced with people who start bandying probabilities about. Like drug companies. Drug companies are very keen on telling you about the risks you face in the world around you. Drug companies are also very keen on selling their products. These two facts may not be unconnected.

For example, statistics suggest that if ten thousand men are not given a thrombolytic agent (no – I don't know what it is either) after a heart attack, then around one thousand will die in the next six weeks. But if the same group of men are given the treatment, then around eight hundred will die in the same period. Without the treatment, the chances of dying are 1000/10 000 or 10/100, and with the treatment, the chances of dying are 8/100. The problem is that this information can be presented in very different ways. The drug companies can say they are saving two hundred lives, or tell you that they are increasing your chances of survival by a fifth. On the other hand, doctors might be concerned that you have to treat one hundred patients in order to save two lives, and feel that this money might be better spent elsewhere on treatments that are more effective. It is a difficult decision to make on a very emotive issue, and it doesn't help that the probabilities can be manipulated to argue for either side.

---

69. Captain Hook has captured you and tied you to the mizzenmast. He lowers a ladder over the side of the boat, and tells you that as soon as the water reaches the thirteenth rung, you are going to be fed to the sharks. You can see that currently the fourth rung is only just submerged, and the Captain informs you that each rung is 5cm wide, that the rungs are 7cm apart, and that the tide is rising at 22cm per hour. How much time do you have left?

It is worth mentioning how probability relates to odds, since they are both talking about the same thing, and odds are what you have to get your head round in the betting shop. If a bookie is offering odds of 2 to 1 on a horse, he is saying that he thinks that the horse will lose two times for every one time it wins. Therefore, he is estimating that the probability that the horse will win is 1/3 and the probability that it will lose is 2/3. Similarly, if the horse is offered at 10 to 1, the bookie reckons that it will lose ten times for every one victory. He reckons that the probability of it winning is 1/11 and the probability of it losing is 10/11. Evens means he reckons the horse is as likely to win as to lose (i.e. the probability of a win is ½ and the probability of a loss is a ½), and odds of two to one on means that he thinks that it will win twice for every one loss (i.e. the probability of a win is 2/3 and the probability of a loss is 1/3). Or at least this is true if he is not trying to make money. Most bookies are keen to try and stack the odds slightly in their favour. After all, they have to put dinner on the table like the rest of us.

---

70. In a hospital, there are three categories of employees: doctors, nurses, and administrators. There are 350 employees at the hospital, and seventy of them are male. Twenty-eight male employees are doctors. There are half as many female doctors as there are male doctors. Twenty-two male employees are in the category of administrators. There are 250 nurses. Out of the 350 employees, one employee is chosen at random to go on a luxury cruise. What is the probability that the person chosen is a female administrator?

So, that is the basis of the theory of probability established. The chance of drawing a red card from a pack of cards is 26/52, the chance of drawing a jack from a pack of cards is 4/52, and the chance of drawing a red jack from a pack of cards is 2/52. But even at this level, probability is already able to help gamblers like Gerolamo Cardano, who are looking to gain an edge in certain games. That is what counting cards in Blackjack is all about.

I thought counting cards must be some kind of cheating, on a level with marking cards, or having the ace of diamonds hidden up your sleeve, or arranging for your girlfriend on the balcony above to use binoculars to read the cards in your opponent's hand, and then radio the information into what you claim to be your hearing aid. But it isn't. In the simplest strategy of counting cards, the gambler just keeps a tally of how many high cards (tens, jacks, queens, kings and aces) have been dealt. The reason for this is that, when there is a high concentration of such cards left in the pack, the odds slightly favour the player against the dealer. This is because, according to the rules of the game, the dealer must take another card if his hand totals a count of less than seventeen, whereas the player is not obliged to do so. When there are many high cards left in the pack, the dealer is likely to go 'bust', by taking cards that total more than twenty-one, whereas the player can be more cautious. As soon as this situation arises, it is time to start betting lots of money.

Now, in my opinion, this is an intelligent strategy – not cheating. But casinos don't like it, and so they make it difficult by giving the dealer several decks of cards to deal from rather than just one, and by kicking you out of their club if they can spot you using card-counting techniques. Which all seems a bit unfair to me, since counting cards when several decks are being used is difficult at the best of times, and impossible after a few martinis that have been shaken and not stirred. On top of that, the vast majority of gamblers aren't even trying to use such techniques.

Even Cardano fell foul of the gambling authorities, despite his attempts to defeat them by the application of mathematics. His indulgence in card games and chess were a constant drain on his resources, so much so that, on one occasion, he lost his sense of perspective and slashed an opponent across the face. I am fairly sure that this sort of behaviour is one of the signs mentioned in those pamphlets that casinos hand out to customers entitled something like: 'How to detect the symptoms of gambling addiction.' Others are 'The belief that you are unbeatable', and 'Total loss of sense of time'.

4 IT'S NOT ALL ABOUT NUMBERS

Probability didn't just appear out of nowhere. At heart, it is a commonsense approach to the fact that the world around us is full of unknowns. We constantly have to reckon the odds in order to make decisions, or resolve problems – just not by using rectangles.

The two men who are normally given credit for being the 'fathers' of probability as a mathematical subject were really only reasoning by commonsense about a problem that had been posed to them. In 1650s Paris, the Duke of Roannez owned a salon in which the leading intellectuals of the day met to discuss important issues and have their wigs blow-dried. It was here that a high-living nobleman Antoine Gombaud (1607-1684), the Chèvalier de Mère, made the acquaintance of a young mathematician by the name of Blaise Pascal.

Born at Clermont in 1623, Pascal first developed an interest in mathematics as a child, when the subject was banned by his father, who was concerned he would be overworked. Giving up his playtime to this new study, in a few weeks he discovered for himself many properties of shapes – in particular the fact that the angles of any triangle add up to 180 degrees. By the age of fourteen, he had been accepted into France's intellectual elite, and was happily writing papers on some of the most taxing problems of the time.

The Chèvalier was a keen gambler, and he asked Pascal several questions connected to his love of dice. Pascal, in turn, mentioned these problems in his letters to another leading mathematician,

Pierre Fermat (1601-1665). The problem that most interested the two men was how they should divide the money at stake if two players of equal skill wanted to stop a gambling game before it had finished. Pascal felt that it should be possible to divide the stakes up fairly to reflect the current state of the game. He discussed this problem in his letters with Fermat, and they came up with different ways of solving it, although they agreed on the correct answer.

Fermat gave his answer in the context of two players playing a game, where the winner is the first to get three points, and each player has put in thirty-two pistoles as a stake.

He first considered what to do, if, at the point they decide to stop, the first player has gained two points, and the second player has only one point. He argued as follows:

If the first player wins the next round, he will win all sixty-four of the pistoles. If he loses the next round, then the two players are on equal scores, and so should share the money.

So, whatever happens in the next round, the first player certainly walks away with thirty-two pistoles. Therefore, he is entitled to put that much money in his wallet now.

With respect to the remaining thirty-two pistoles, the first player has a 1 in 2 chance of winning it (since both players are equally skilled). Therefore, if the game is stopped, he is entitled to take half of it. In total, then, the first player should take forty-eight pistoles, and the second player sixteen.

Pascal then considered how to divide the stakes if the first player has scored two points, and the second player has scored no points (with the winner still being the first to three, and both players having staked thirty-two pistoles as before). In this case, he argues:

If the first player loses the next round, the two players are in the situation that he has just dealt with, namely that the first player will have two points, and the second player will have one point. For this situation, he has already explained why the first player can claim 48 pistoles and the second player can claim sixteen pistoles.

If the first player wins the next round, he will win the game, and can claim all sixty-four pistoles.

Therefore, using similar reasoning to the above, the first player can say that, in the worst case scenario, he will win forty-eight pistoles – and so that amount of money is definitely his. With respect to the remaining sixteen pistoles, he is as likely to win them as to lose them, and so he is entitled to take eight of them.

Therefore, if the game is stopped with the score being 2–0, the first player can happily pocket fifty-six pistoles, and the second player can go and drown his sorrows with the remaining eight.

---

71. Pascal then deals with the situation where the first player has scored one point and the second player has scored no points. Give it a go – this is your chance to claim that you too could have been a mathematical genius, if only you had been born in Paris during the 1650s. (Ah, the twists of fate that shape our destinies…) Remember that the first player to get three points is the winner, and each player has staked thirty-two pistoles.

To move away from all this gambling, it might be useful to next look at a couple of different tools that can help us deal with problems involving probability. The first of these is called a probability space diagram, which is simply a method of making sure that you have thought of all the possibilities in a given situation. If we go back to the problem that d'Alembert got wrong – about the chance of getting at least one head when flipping a coin two times – it is possible to set out all the possibilities using a probability space diagram like the one below:

	second throw	
	heads	tails
first throw — heads	HH	HT
first throw — tails	TH	TT

Using a diagram like this makes it impossible to miss out any of the possible outcomes for throwing a coin twice. There are four possibilities, three of which involve throwing at least one head. Therefore, the probability of getting at least one head in two flips of a coin is ¾.

A similar diagram can be used whenever you are faced with a problem where one event is followed by another event, and you want to work out all the different possibilities for the two events combined. For example, let us say that you are locked in battle with your sister over a game of Monopoly (please excuse the British bias here). You have just landed on Piccadilly, you already own Coventry Street, and Leicester Square is still up for grabs. You are very tempted to buy, but your funds are low and your sister has three houses on both Mayfair and Park Lane. If you buy Piccadilly, you will not have enough ready cash to cover your costs if you were to land on your sister's properties, which you will do if you roll an eight or a ten. You really want to win.

second dice

	1	2	3	4	5	6
1	11	12	13	14	15	16
2	21	22	23	24	25	26
3	31	32	33	34	35	36
4	41	42	43	44	45	46
5	51	52	53	54	55	56
6	61	62	63	64	65	66

first dice (label at left of table)

In this situation, you can work out what the chances are of landing on Mayfair and Park Lane using a diagram similar to the one above, because you are combining the score on two dice.

From the table, you can see that there are thirty-six different possible outcomes when you throw two dice, although some of them give the same score. (Rolling a three and then a four is different from rolling a four and then a three.) Of these thirty-six, there are five ways of scoring eight, and three ways of scoring a ten. So the chance of landing on Park Lane or Mayfair, and being forced to part with large wads of fake cash, is 8 in 36. You would be pretty unlucky to suffer at the hands of a gloating sister. Maybe it is worth buying Piccadilly.

> 72. A man left his home to go hunting and travelled due south. He spotted a bear moving due east. The hunter tracked it for a few kilometres until he was close enough to shoot and kill it. He then headed due north back home. What colour was the bear?

A tree diagram is another tool that is often used to deal with problems involving probability. I have no understanding of whether a tie goes with a shirt, or a shirt goes with a pair of trousers, or a pair of trousers goes with a tie, and so, when I ponder about what to wear to work, I am not limited by questions of taste. I am a truly free agent. In this situation, a tree diagram can be used to represent all the different combinations present in my wardrobe, which is made up of the following items: a blue shirt, a green shirt, a red shirt, black trousers, grey trousers, a yellow tie, and a pink tie.

From the tree diagram on the following page, it is possible to see that I have twelve options available to me when it comes to dressing for work, although I suspect that more than half of them will be aesthetically unsatisfactory. Each path through the tree diagram from left to right gives one of these options. We can see that there is a one in twelve chance of me choosing the first outfit for example, which is a combination involving my red shirt with my pair of black trousers and the yellow tie.

Probability space diagrams and tree diagrams are both strategies for making sure that you are aware of all the different possibilities in a given situation. However, they can be adapted to deal with more complicated probability problems, as you will see in the next chapter.

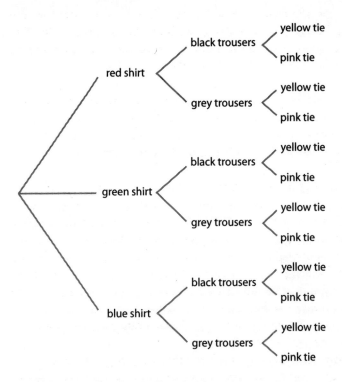

73. You enter a town that has 101 killers and 101 pacifists. When a pacifist meets a pacifist, nothing happens. When a pacifist meets a killer, the pacifist is killed. When two killers meet, both die. Assume meetings always occur between exactly two persons, the pairs involved are completely random, and that you have to walk the streets of the town as if there is not much to worry about. What are your chances of survival?

5 THE WEATHER FORECAST IS WRONG

Mathematicians in general tend to underestimate just how hard mathematics is for other people. Fermat, the man who helped Pascal invent the science of probability, wasted hours of other peoples' time by writing his proofs in crabbed hand-writing in the margins of books, so that they were very difficult to follow. Worse than that, he was very fond of missing out large chunks of his argument, simply saying that it was easy to see how one result followed from another. His refusal to lay out his work in a sensible manner led to his 'last theorem'. In his copy of a book by the ancient Greek mathematician, Diophantus, he scrawled a note saying that he had found a proof for the fact that the equation $xn + yn = zn$ has no integer (whole number) solutions, when n is greater than 2, but that there was not room enough to write it. The search to find this proof lasted until 1995 – 350 years of toil just because he wasn't prepared to get out a piece of scrap paper. Some of Fermat's disgruntled friends pointed out that leaving these gaps in his argument was a very annoying habit. Fermat had to admit that they were right, when he had serious trouble trying to work out just how he had proved some of his results.

So it is a real pleasure when a mathematician puts his hands up and acknowledges that what he is about to say is a bit difficult to follow. Laplace was one such honest man. 'One of the most important points of the theory of probabilities and that which tends the most to illusions is the manner in which probabilities increase or diminish by their mutual combination.' (*Théorie Analytique des Probabilités*, 1814). He is talking about the difficulty of working out what the probability of a series of events happening is, if you know the probability of each individual event happening. For example, what is the probability that you roll a six on a dice, then flip a coin and get heads, and then select a red jack from a pack of cards?

This issue has caused havoc over the years, and still leads to all kinds of confusion. For example, I recently saw a television weatherman making his forecast at the weekend. He confidently stated that as there was a 50% chance of rain on the Saturday and there was also a 50% chance of rain on the Sunday, we were certain to get some rain over the weekend. In other words, he added together the two probabilities to get a 100% chance.

But this can't be the case. Firstly, nothing is ever certain. And secondly, imagine that he was making a three-day forecast, and that on the Monday there was also a 50% chance of rain. By his logic, there would be a 150% chance of rain over the three-day period. But this makes no sense. What could a 150% chance be? How can anything be more certain than a 100% probability?

---

74. In Rwanda, it is necessary to go on a two hour bus journey from the capital, Kigali, to Ruhengeri in order to go and see the mountain gorillas. Virunga buses operates this route, and they have two deluxe coaches, and two dilapidated mini-buses. The buses leave hourly from both Kigali and Ruhengeri, and turn around immediately on their arrival at their destination to make the return journey. The first bus is at 7.00 in the morning and the last bus is at 6.00 in the evening. The two luxury coaches both start at 7.00 (one from Kigali and one from Ruhengeri). If you turn up to get on a bus at random, what are the chances that you will get to go in a deluxe coach?

The thing to remember is that when you are combining probabilities you don't add, you multiply. Yet it is not at all easy to see why this might be. To return to the weather problem, and to deal with it logically, it is useful to draw a tree diagram.

In this case, the diagram can be used not only to list all four possibilities for the weather for the weekend, but also to record the probabilities involved in the problem:

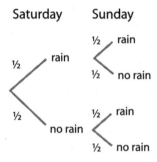

Now, what the weather forecaster is saying when he says that there is a 50% (or 1 in 2) chance of rain falling on Saturday, is that, given one hundred Saturdays with exactly the same weather characteristics, it would rain on fifty of them, and not rain on fifty of them (roughly – even the most trusted weather forecasters must admit that meteorology is not an exact science). So, let us pretend that we are caught in an eternally repeating week, in which the weather patterns are always exactly the same. After we had lived through one hundred of these weeks, we would expect to have experienced fifty rainy Saturdays, and fifty rain-free Saturdays.

Take the fifty rainy Saturdays. Each of these will certainly be followed by a Sunday. Now focus on the fifty Sundays that follow the fifty rainy Saturdays. Assuming that the weather on Saturday has no effect on the weather on Sunday (which is probably not true, but let it go), on each of these fifty Sundays, the weather forecaster has predicted that there is a 50% chance of rain. So, we expect that, of these fifty Sundays, half of them (i.e. twenty-five) will have rain, and half of them (i.e. twenty-five) will not. Therefore, there are twenty-five weekends where it rains on both Saturday and Sunday, and twenty-five weekends where it rains on the Saturday but not the Sunday.

Similarly for the fifty weekends where it did not rain on the

Saturday. Each of these Saturdays will also be followed by a Sunday, and each of these Sundays has a 50% chance of rain. Therefore, of the fifty Sundays that follow rain-free Saturdays, we would expect half (i.e. twenty-five) of them to be wet, and half (i.e. twenty-five) of them to be dry. So, there will be twenty-five weekends where it is dry on Saturday but wet on Sunday, and twenty-five weekends where it is dry on both Saturday and Sunday.

Therefore, overall, out of the one hundred weekends that we have sat through, listening to the same football scores, watching the same programmes, and telling the same jokes, twenty-five of them will have both a rainy Saturday and Sunday, twenty-five will have a rainy Saturday and a dry Sunday, twenty-five will have a dry Saturday and a rainy Sunday, and twenty-five will have both a dry Saturday and a dry Sunday. In other words, we expect each of these possibilities twenty-five times out of one hundred, which means that the probability of each is ¼. Here is the same tree diagram, with this new information added in:

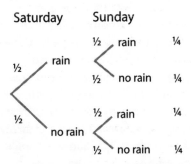

First of all, you can see that in order to find the probability of two events happening one after the other, you multiply together the probabilities of the two events happening on their own. For example, the probability that it will rain on Saturday and on Sunday is ¼. This is the result of multiplying together the probability that it will rain on Saturday (1/2) with the probability that it will rain on Sunday (1/2). ½ × ½ = ¼.

The explanation for this follows from what has been said

above. We have already seen why, if you take one hundred weekends where the probability of rain on the Saturday is a ½ and the probability of rain on the Sunday is also a ½, you expect rain to fall on both days for ½ × (½ × 100), or twenty-five, of the weekends. But the same argument applies to whatever number of such weekends we decide to look at. If we took twenty weekends, we would expect rain to fall on both days for ½ × (½ × 20), or five, of them. If we took five thousand weekends, we would expect rain to fall on both days for ½ × (½ × 5000), or 1250, of them. In each case, the probabilities are multiplied together, so that we can say that, in general, the chance of rain falling on both days is ½ × ½.

Secondly, it is clear that the weather forecaster is wrong. Because there is a 1 in 4 chance that it will stay dry on both days of the weekend. He should have announced that there was a 75% chance of rain on at least one of the days of the weekend, since this is true for three of the four weather possibilities over the weekend, and each of these possibilities has a 25% chance of occurring.

Still, this forecaster is among good company. Plenty of mathematicians have made exactly the same mistake. Mr Cardan, for example, was unable to calculate how many times you have to throw one dice before you have an even chance of throwing at least one six on one of the throws. Or rather, he did calculate an answer, but it was wrong. He thought that, since the chance of throwing a six in one throw is 1/6, the chance of throwing at least one six in two throws is 2/6, and the chance of throwing at least one six in three throws is 3/6 (or ½).

He has made exactly the same mistake as the forecaster. There is no way that he can be right, since, by his reasoning, after six throws of the dice, you will have a 6/6 chance of rolling at least one six. In other words, Cardano is claiming that you will certainly throw at least one six, if you roll a die six times. However, experience tells us this is not true, since it is definitely possible to fail to get a six in six throws of a die.

6 BACK TO THE CLASSROOM

Now that probability space diagrams and tree diagrams and combined probabilities are no longer a mystery, it is possible to return to some of those problems about sweets and marbles and dice and balls in bags that textbooks are so obsessed with.

After several rounds of interviews for three posts, a company has narrowed down the candidates to a group of four: three women and a man. However, they are unable to find any way of choosing between them, and so they decide to put their names in a hat, draw out three names, and select them in this way. What is the probability that the man will be one of those selected?

In any probability problem, the first thing to do is to work out the different possibilities in the given situation, using whatever technique is the most suitable. In this case, the simplest method is simply to list the different possibilities for the three names that come out of the hat. If you call the candidates, W_1, W_2, W_3, and M, then there are only four possibilities for who ends up employed: W_1, W_2, W_3, or W_1, W_2, M or W_1, W_3, M or W_2, W_3, M. (It does not matter in what order the names are selected from the hat). Since these four possibilities are equally likely, and since in three of them the man is employed, the probability of the man being selected is ¾.

It is perfectly possible to use a tree diagram to solve the problem. It starts off something like this:

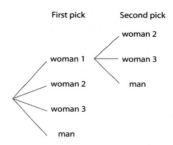

However, for this problem, you can see that this approach is going to be far more complicated than simple listing. It will get you there in the end, but only after you have drawn a lot of branches.

It is time to return to issues involving dice. In a particular board-game, two dice are thrown, and the difference between them is calculated. What is the probability of getting a difference of zero? What is the probability of getting a difference of zero or two? What is the probability of getting a difference that is either an even number or a square number?

As before, the first thing to do is to work out all the different possibilities in this scenario. You could simply list them as before, but since there are quite a few of them, you run the risk of missing some out. You could use a tree diagram, but, as before, it is going to contain a disturbing number of branches. The best tool for this problem is the probability space diagram:

second dice

	1	2	3	4	5	6
1	2	3	4	5	6	7
2	3	4	5	6	7	8
3	4	5	6	7	8	9
4	5	6	7	8	9	10
5	6	7	8	9	10	11
6	7	8	9	10	11	12

first dice

The probability space diagram ensures that no possibility is missed out, and enables the difference between the numbers on the two dice to be recorded. You can see that there are thirty-six different possibilities, although many of the combinations of the numbers on the two dice result in the same difference (e.g. a 4 on the first dice and a 3 on the second gives the same difference as a 5 on the first dice and a 6 on the second).

Since each possibility shown in the diagram is equally likely, the required probabilities can now be calculated. There are six ways of getting a difference on the two dice of zero, and so the probability of this event is 6/36.

There are eight ways of getting a difference of two on the two dice (i.e. the probability of getting a difference of 2 is 8/36). Therefore, there are 6 + 8 (or 14) ways of getting a difference of zero or two on the two dice, and the probability of this event is 14/36. In this case, the probability of getting a difference of two or zero is equivalent to the probability of getting a two plus the probability of getting a zero (14/36 = 6/36 + 8/36). The reason for this is that it is not possible for both conditions to be satisfied at the same time. There is no combination on the two dice that will give both a difference of two and a difference of zero. If this is the case, the two options are called mutually exclusive.

You have to be a bit careful when answering the third question. The probability of getting a difference that is an even number is the same as the probability of getting a difference of two or four, and this can be calculated to be 12/36. The probability of getting a difference that is a square number is the same as the probability of getting a difference of one or four, and this can be calculated as 14/36. However, you cannot find the probability of getting a difference that is either an even number or a square number by simply adding these two probabilities, because four is both an even number and a square number. If you just add the two probabilities, you will double-count the combinations on the dice that give four as a difference. In this situation, the two options are not mutually exclusive. It is possible to find combinations of the two dice that give both an even-number difference and a square-number difference. In fact, if you simply highlight the different possibilities in the space diagram that lead to an even-number difference or a square-number difference, you find that there are twenty-two of them, and so the probability of getting such a difference is 22/36:

second dice

	1	2	3	4	5	6
1	0	1	2	3	4	5
2	1	0	1	2	3	4
3	2	1	0	1	2	3
4	3	2	1	0	1	2
5	4	3	2	1	0	1
6	5	4	3	2	1	0

first dice

---

75. Six men and twelve women are queuing at the hairdresser. Half of the men have grey hair and so do half of the women. What are the chances that a person chosen at random will be either a man, or grey-haired or both?

And so to balls in bags. A bag contains three red balls and five blue balls. You select one at random, note its colour, and return it to the bag. You then select another at random, note its colour and return it. What are the chances that the two balls you select are red? What are the chances that the two balls you select are blue and red (in any order)?

As always, the first task is to work out the different possibilities for the two balls. There are various options open to you. You could label the red balls R1, R2, and R3, and the blue balls B1, B2, B3, B4, and B5, and then list all the possible combinations of two of these balls, or draw a probability space diagram (as in the last problem) or a tree diagram. However, since you are only concerned with the colour of the balls, such an approach will lead to a mass of superfluous information. You don't need to know exactly which of the balls are selected, just what there colours are.

As a result, you can simplify the task. You can still use one of several approaches. You could list the possible colour combinations (RR, RB, BR, BB) or display them in a probability space diagram:

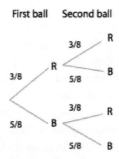

Or show them in a tree diagram:

You can select whichever method you like. However, it is important to realise that, since there are not the same number of red balls and blue balls in the bag, the different possibilities are not equally likely. The problem is more complicated than the previous ones. In fact, the chance of selecting a red ball at random from the bag is 3/8 and the chance of selecting a blue ball is 5/8. This information has been represented on the tree diagram above.

Once this has been done, you can solve the problem, because you know that when you find the probability of one event followed by another, you multiply the respective probabilities. Therefore, the probability of removing two red balls is 3/8 × 3/8, or 9/64. The probability of selecting a blue ball and a red ball in either order is the sum of the probability of selecting a red and then a blue and the probability of selecting a blue and then a red: $(3/8 \times 5/8) + (5/8 \times 3/8)$, or 30/64.

There is one further complication, and that occurs when you do not replace the first ball after you have noted its colour. In the above problem, the first event (selecting the first ball) has no affect on the second event (selecting the second ball). In this case, the two events are said to be independent. However, if you do not replace the first ball, then the first selection of the ball does affect the selection of the second ball. If you pull out a red ball on the first selection, then there are only two reds and five blues left in the bag. The chances of getting a red on the second selection are 2/7, and the chances of getting a blue on the second selection are 5/7. On the other hand, if you pull out a blue ball on the first selection, the chances of getting a red on the second selection are 3/7, and the chances of getting a blue on the second selection are 4/7. The colour of the ball from the first selection affects the probabilities for the second selection. In this type of situation, the two events are called conditional.

However, whatever names you give to the different situations, you can still solve the problem in the same way. If you do not replace the first ball, then the tree diagram looks like this:

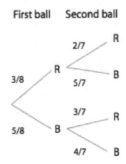

The probability of selecting two reds is now given by $3/8 \times 2/7$ (or 6/56), and the probability of getting a red and a blue in either order is: $(3/8 \times 5/7) + (5/8 \times 3/7)$, which turns out to be 30/56.

76. A bag contains ten beans. Seven of the beans are black and the rest of the beans are red. One bean is taken out and its colour is observed. It is not replaced. A second bean is removed and its colour is observed. Find the probability of picking two beans of the same colour.

Outside, shadows are claiming the shabby cars of the teachers, and concealing the discarded sweet wrappers on the ground. One or two sixth-formers are already slipping away from the compound, as the end of the day draws nearer.

The class is full of artificial yellow light, as if the central powers that live in the offices down the forbidden corridors wish to keep fatigue at bay with pure brightness. The clock shows that there are ten minutes left. Every single student (with the possible exception of Bernadette) wills the hands of the clock towards their final destination, whilst ears strain to catch the noise of other classes being let out early.

The remaining minutes seem so futile. They will eventually pass. Nothing will be achieved in them. So what is the point in living through them? Charlie fantasises about special mental powers that allow him to fast-forward time.

Mr Barton is just as aware of the clock as his prisoners. He does not understand why young people are so happy to waste valuable minutes in the classroom. If they do not practice doing the same type of question over and over again, how will they ever memorise the methods involved in solving them? He stares in annoyance at Charlie's legs, which are drumming up and down beneath his desk, as the last coke from lunch releases its uncontrollable energy. In response to his glare, Charlie tries hard to sit still.

There are nine minutes remaining.

7 PUTTING PROBABILITY INTO PRACTICE

At this stage you may be beginning to see how probability can be really helpful in all sorts of ways. In particular, it can help you to do battle on equal terms with those irritating people that always seem to win any kind of game that they play. In Monopoly, they lend you money at extortionate rates of interest, whilst they construct massive leisure developments on Mayfair. In Scrabble, they use their knowledge of obscure two-letter words from Oriental religion to position 'Q' on a triple-letter score.

There is a dice game called Perudo. Everyone starts off with six dice, which they roll and keep hidden under a cup. Only they are allowed to see their own dice. It is a bidding game, and so someone starts by making a call of e.g. 'three twos'. This means that he (or she) believes that there are at least three twos amongst the combined dice of all the players. (There is a further complication involving 'ones' which I will come to in a while). The following person may call three of a number higher than two, e.g. 'three threes' or 'three fives', or four (or higher) of any number they like (e.g. 'four twos', 'four fours', 'five threes', 'ten fives' etc.). The bid goes from one person to the next until there comes a stage when someone thinks that the person who has just bid is mistaken, and that there are not e.g. 'eleven fives' amongst all the dice of all the players. At this point, he calls the bid. If there are eleven fives, he loses a die. If there are not eleven fives, the bidder loses a die. The aim of the game is to be the last person left with any dice.

People who always win at games love Perudo, because there

is a strategy behind the bids you should make, especially at the beginning of the game, when there are a lot of dice around. You might make a bid, and notice that your neighbour has raised an eyebrow in surprise, or chuckled knowingly to themselves, or made a gentle 'tutting' sound underneath their breath. This is because you have not made your bid according to the strategy. Of course, you can refuse to stoop to their levels, and carry on making your bids as the whim takes you. But, just in case you no longer want to be on the receiving end of such disdain, it might be worth knowing that Perudo relies on probability.

Say there are six people at the start of the game, and each person has six dice. Then there are thirty-six dice in total. It is your turn to make the first bid, and you want to figure out a sensible call. Five is your lucky number, but the question is how many fives do you bid? Probability says that roughly one die in every six will show a five, which means that you would expect there to be around six fives in total amongst the thirty-six dice. Therefore, a good starting call would be 'six fives'. The person next to you is unlikely to challenge you, and the bidding is unlikely to ever get back to you, because the bids will be forced improbably high before it does. In fact, since the same argument applies to whatever number you decide to call; six of anything is a good bid to start off with.

Now it is later in the game. Eight dice have been lost (so there are only twenty-eight dice left), and the person before you has just bid 'six threes'. Should you challenge him, or should you make a bid of your own? As before, probability says that roughly one in every six dice will be a 'three', and so, out of twenty-eight dice, you would expect four or five of them to show 'threes' (since $28 \div 6 = 4.6$ recurring). It is a difficult decision to make. If you play strictly according to probability, you should call him. However, you might want to take into account what dice you have, and your knowledge of the kind of player he is, before you make your mind up. I didn't say probability would tell you what to do – just help you along the way.

There is a further complication in Perudo that needs to be

taken into account. Ones are wild and count as whatever number as being bid at the time. Therefore, when someone makes a call of 'ten fives', they are actually saying that they think there are at least ten fives AND ONES amongst the combined dice of all the players. And if the next player goes on to bid 'eleven twos', they reckon that there are at least eleven twos AND ONES.

Probability can deal with this. The situation is the same. There are six players with six dice, and it is your turn to start the bidding. Five is still your lucky number. You reason that roughly one in every six dice will be a 'five', and that roughly one in every six dice will be a 'one'. So, out of the thirty-six dice around the table, you expect there to be approximately six 'fives' and six 'ones'. In other words, you reckon that there are twelve dice in total that will count as 'fives'. Your starting call is 'twelve fives'.

Later in the game, after fourteen dice have been lost (leaving twenty-two remaining), your eyebrow-raising neighbour looks at you and without a flicker of emotion says: 'ten sixes'. You return his gaze without concern. Just as before, you expect one in six of the dice to be a 'six', and one in six of the dice to be a 'one'. Therefore, there should be around three or four 'sixes' and three or four 'ones' amongst the dice. Your neighbour's call seems a bit ambitious, especially as you do not have any 'sixes' or 'ones' amongst your own dice. What is more, your neighbour can sense that you know all this, and suddenly his face is no longer emotionless – there is fear in his eyes. You call him, and everyone reveals their dice. There are only seven 'sixes'. You watch with satisfaction, as he hurls his die angrily across the room, and storms out, muttering something about 'people who take the game too seriously'.

There you go. That is the strategy that lay behind his looks, and raised eyebrows, and 'tutting', and now you can make use of it to silence him forever. All you need to do is to keep a track of how many dice are left in total as the game progresses, and use probability to make a sensible guess at how many dice of each number there are likely to be at any given time.

However, bear in mind that probability will not always protect you. Sometimes there will be an improbably high or improbably low number of dice showing a particular number, especially when the total number of dice is low, but, if you stick to the strategy, you are likely to hold on to your dice. Then you, in your turn, will be able to raise an eyebrow at some unfortunate who is not taking the laws of probability into account. Or perhaps you will choose to maintain a dignified silence...

---

77. a) You are first to bid in a game of Perudo. You have been counting the dice that other players have lost, and you know that there are twenty-four dice around the table. What is a good starting bid?

b) Your neighbour has turned to you with a smirk flickering at the corners of his mouth. He has called six threes. There are twenty dice around the table, including your own (you have five dice left). You have two threes and no ones amongst your dice. What do you do?

c) Your neighbour has called again: seven fives. There are nineteen dice left, of which four are yours. You have one five, and no ones. What do you say?

---

8 VEGAS, BABY!

I am not a gambling man, but when you find yourself in Las Vegas, there is not much else to do. Of course, there are different ways to go about your gambling. You can eke out your chips by betting them one by one on the Blackjack tables. If you really want to spin it out, it is best to pass on a few hands in the pretence that you are thinking deeply about your strategy – just don't let the casino confuse you with a card-counter. The advantages to this approach are that you don't walk away from the table empty-handed after five minutes, and that since you are in Las Vegas (where anyone gambling can lay claim to free drinks), you can make up for the loss of your cash with a series of ever-more-unpleasant cocktails.

The disadvantage to this approach is that you don't look like a PLAYER. Far better to buy one gleaming multi-coloured chip for one hundred dollars, and gamble it all on one spin of the roulette wheel. Far better because the adrenalin rush is spectacular. And far better because, in terms of probability, this is the most sensible way to play roulette. Gamble big on the first spin, and then walk away, whatever the result. I walked away with a big grin and two hundred dollars, which I promptly spent on bottles of champagne for me and my friends, and entry to a special kind of club.

In order to work out why this approach to roulette not only appears the most gutsy, but is also, in fact, the most sensible, it is necessary to look at probabilities, and how to use them to calculate how much money you would expect to win from a gambling game. Imagine a game played on two dice. If you roll a double, you win two pounds, except if you roll a double six, in which case you win five pounds. If you roll any other combination, you lose a pound. You want to know whether it is worth your while playing the game.

In order to work out the probabilities of each of the three possible outcomes from one throw of the dice, it is necessary to draw a probability space diagram:

second dice

	1	2	3	4	5	6
1	0	1	2	3	4	5
2	1	0	1	2	3	4
3	2	1	0	1	2	3
4	3	2	1	0	1	2
5	4	3	2	1	0	1
6	5	4	3	2	1	0

first dice

From the diagram, you can see that there are thirty-six different possible combinations when you roll two dice. So, with reference to the game you are playing, you have only a 1 in 36 chance of rolling a double six, and winning five pounds. You have a 5 in 36 chance of rolling a double other than a double six, and winning two pounds. And you have a 30 in 36 chance of failing to roll a double and losing a pound. This information is shown in the tree diagram below:

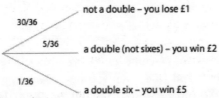

30/36 — not a double – you lose £1

5/36 — a double (not sixes) – you win £2

1/36 — a double six – you win £5

From here, it is possible to work out how much you should expect to win or lose at the game. According to the probabilities, if you were to play the game thirty-six times, you would expect to lose a pound thirty times, win two pounds five times, and win five pounds once. So overall, after thirty-six games, you would expect to have:

$$(5 \times 2) + (1 \times 5) - (30 \times 1) = £\text{-}15.$$

If you played this game thirty-six times, you should expect to lose fifteen pounds. So, on average, for each time you play, you can expect to lose 15/36 pounds, or nearly 42p. This figure is called the expected winnings in this game. Although, in this case, it is actually the expected losings.

Now, it is possible to see why betting once and big at roulette is the best way forward. There is a difference between roulette wheels in the US and the UK. For what follows, I will refer to a US roulette wheel, but the same type of argument applies to different formats. The US roulette wheel has the thirty-six numbers, 1–36, and also 0 and 00. Eighteen of the numbers are red, eighteen of them are black, and the two zeros are some other colour. You step up to play, place your brightly-coloured chip to show that you think that the little white ball will end up on a black number, and step back. The ball hurtles crazily around the wheel, ricocheting off the little metal pins, whilst your heartbeat increases to unsafe levels. It is just you, the white ball, and the spinning numbers. Here are your chances:

If you play thirty-eight times, you should expect to win eighteen times and lose twenty times. Overall, you win 18 × 100, or $1800, and you lose, 20 × 100, or $2000. In other words, after your marathon session on the roulette wheel, you have lost $200. On average, for each spin of the wheel, you are throwing away 200/38, or $5.26 per spin.

But everyone knows that gambling is addictive. If you lose, the demons in your head quickly begin to whisper that you can win it all back if you play just one more time. Ignore them, don't do it, walk away. Because if you play twice, using the strategy shown in the diagram below, these are your possibilities:

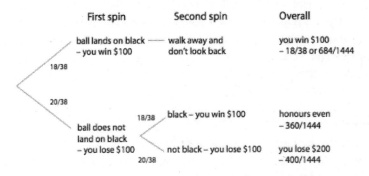

The overall probabilities have been calculated by multiplying along each branch of the tree diagram, since you know that that is how you work out the probability of combined events. For example, the chances of not getting a black on the first roll and getting a black on the second roll are found by multiplying the chance of not getting a black (20/38) with the chance of getting a black (18/38).

By this stage things are getting a little complicated, but imagine that you visited the casino for 1444 nights in a row, and each time played according to the strategy shown in the diagram. Then over the 1444 nights, you would expect to win $100 on 684 occasions, and to lose $200 on 400 occasions. The rest of the time you would expect to walk away with as much money as you walked in with. So, over this period of gambling (which is almost eight years of your life – are you sure it is worth it? – think of the wife and kids), you will have won 684 × $100, or $68 400, and lost 400 × $200, or $80 000. In other words, you will have made a total loss of $11 600. If you average this loss out over the 1444 visits, it works out at around $8.03 per night.

Even if you try and play with the system, by betting more

money on the second throw, or by offering silent prayers to the gods of luck, your expected loss is never as little as if you just play once. The more times you play, the more likely that the odds in favour of the casino will catch up with you. It's a mug's game.

---

78. In a class of sixty pupils, forty-eight pupils like flower-arranging, fifty-two pupils like brass-rubbing, and forty-two pupils like both subjects. Find the probability that a pupil likes at least one of the two subjects.

There is one system, called the Martingale system, which is particularly tempting. You bet £10 on the first spin of the roulette wheel. You walk away if you win, and bet double as much (i.e. £20) on a second spin if you lose. The idea is that if you win on the second spin, you will cover your initial loss of £10, and make a further £10. You walk away if you win on the second spin, but if you lose, you bet double as much again (i.e. £40) on a third spin. Once again, the idea is that if you win on this spin, you will cover your losses so far (£10 + £20 = £30), and make an additional £10. If you lose again, you double your stake once more, and you keep on following this strategy until you eventually win. As soon as you do, you will cover all your losses up to this point, and make £10. At this point, you can start the whole process again. The thinking is that it is very unlikely, almost impossible, that you can play, for example, fifty times, and fail to win once. Easy money…

It is a very persuasive argument, but fatally flawed. It does not take much of a losing streak before you need huge amounts of money to cover your losses so far. For example, if you lost ten times, you would have to put down £10 240 to try and get your money back. I would personally feel a bit nervous about getting involved with this kind of money in an attempt to make a measly £10 against the casino, and even if you are made of stronger stuff, casinos, in general, have a limit on how much money you are allowed to place on a single bet. I think, perhaps, it is better to stay at home.

Not that staying at home is necessarily a safe option. You can gamble at home too. There is always Russian roulette. All you need is a loaded gun and a couple of disposable friends.

---

79. You are in a game of Russian roulette, but this time the gun (a six shooter revolver) has three bullets in a row, in three of the chambers. The barrel is spun only once. Each player then points the gun at his (her) head and pulls the trigger. If he (she) is still alive, the gun is passed to the other player who then points it at his (her) own head and pulls the trigger. The game stops when one player dies. Now to the point: would you rather be first or second to shoot?

Of course, life itself is just a game. It differs from the games found in a casino only in that its rules are infinitely more complicated, and there is no instruction manual to tell you what they are. So, it is unsurprising that people have tried to apply the logic of expected winnings to it.

Blaise Pascal was a very religious man. Early in his career, he decided to give up mathematics, because he felt that it got in the way of his devotion to God. He only returned to it, when, in an attempt to distract himself from the pain of a toothache, he considered various geometrical problems. Miraculously, the pain did not just lessen, it went away altogether. He took this as a sign that God approved of mathematics, and returned to his study of it.

However, he still remained influenced by his religious beliefs. In particular, he wanted to convince people of the importance of keeping the faith, and he was prepared to use any argument in defence of this belief. He made one such attempt using the concepts of probability that he had helped to discover. He argued that you could choose either to live according to religious values or not. If you did not, you might be able to indulge yourself with all sorts of pleasures and luxuries during your mortal life, but you stand no chance of attaining the everlasting bliss of heaven. If you do live religiously, you may miss out on things like double cream and orgies, but even if the probability that God exists is small, you have given yourself a shot at eternal happiness. Pascal argued that the expected gain of a measly lifetime of self-gratification, is nothing compared to even a small chance of going to heaven. Therefore, if you accept the laws of probability, you should lead a religious life.

Just in case, you are patting yourself on the back for having made the right choice, it is not enough just to turn up to Midnight Mass once a year and bawl out wine-fuelled carols. The locks on the gates of Heaven are a little more complicated than that.

9 THE LAW OF LARGE NUMBERS

Probability can't tell you for sure what is going to happen, but it can tell you what is likely to happen, and that isn't bad. As British philosopher John Locke (1632-1704), author of the monumental *An Essay Concerning Human Understanding* glumly pointed out: '...in the greatest part of our concernment, God has afforded only the Twilight, as I may so say, of Probability, suitable, I presume, to that state of Mediocrity and Probationship He has been pleased to place us in here.'★

So far, you have mostly been looking at how to apply probability to different scenarios involving games of one kind or another. You have defeated your younger sister at Monopoly. You have cleaned up in Vegas. And you have cheated at Blackjack. But, as Locke says, probability is part of the everyday make-up of human life. It would be very useful if you could apply probability to situations that do not involve rolling dice or picking cards.

However, this creates a bit of a problem. As mentioned previously, the reason why probability is most easily applied to game situations is that they have very clear rules, and that these rules are designed for fairness. People would not play games if they thought that they did not start off with as good a chance of winning as anyone else. You would not play backgammon if you knew that your dice were loaded so that they would only ever show a one or a two.

This means that the number of situations where you can actually work out the probability of something happening beforehand is

★ Locke (1997) p576.

very limited. You can make fairly accurate statements about the probability of rolling an odd number of a die (although even a statement like this contains many assumptions e.g. that all the dices faces are exactly equal), but, in general, most situations are too complex to be able to analyse in this way. You have already seen that it is not possible to put a number on the probability of a particular human action achieving its intended aim, and, in fact, I suspect life would become rather depressing if it were. It would contain none of its little surprises, if you could say with absolute certainty that the chances that you would meet your life partner at the bowling alley this very evening were four in five. There would be no heart-stopping moment of recognition, no swell of imaginary orchestral music, no erratic beating of your heart.

But, if we leave it there, then probability isn't all that helpful. So it is very lucky that a man called Jacob Bernoulli didn't give up hope. Born in Switzerland in 1654, Bernoulli was one of the leading mathematicians of the age, and one of the first to try and understand the complexities of Leibniz's newly-discovered calculus (which is A-Level stuff nowadays, although most of us don't pretend to understand the theory behind it). Bernoulli spent most of his life in fierce rivalry with his equally talented brother Johann. If you want to maintain family harmony, it is best for brothers to follow different paths in life, and it was clear that these two brothers were particularly competitive. However, both chose to become mathematicians, and the result was years of one-upmanship, slanging matches, and uncomfortable family Sunday lunches.

---

80. A stranger stops you in the street and tells you that he has two children, at least one of which is a girl. What are the chances that he has two daughters?

In the field of statistics and probability, Jacob Bernoulli's most important contribution was his Law of Large Numbers. This law states that 'the more trials made the closer the ratio of success to trials approaches the value of theoretical probability.' For example, you know that the chances of rolling a five on a fair die are 1 in 6. This does not mean that you will definitely throw a five, every time you roll the die six times, but, according to Bernoulli's Law, it does mean that the more times you throw the die, the closer the ratio of the number of fives thrown to the number of rolls of the dice comes to the theoretical ratio of 1 in 6.

So, it is quite possible that in twelve throws of the dice (i.e. not very many), you will get a strangely high number of fives, or none at all, rather than the two predicted by the theoretical probability. In this case, what actually happens is very different from what theory says should happen. If you rolled four fives, then the ratio of fives to the number of throws of the dice would be 4 in 12, or 1 in 3, which is very different from 1 in 6. And if you rolled only one five, the ratio of fives to the number of throws would be 1 in 12 – again, a long way off from the theoretical 1 in 6.

But if you roll the dice 120 times, the ratio of the number of fives to the number of throws is much more likely to be close to 1 in 6. It is unlikely that you will get exactly twenty fives (the number of fives that the theory predicts), but you are very likely to roll a number of fives that is fairly close to this total – say between fifteen and twenty-five fives. If you rolled fifteen fives, the ratio of fives to throws would be 15 to 120, or 1 to 8, and if you rolled twenty-five fives, the ratio of fives to throws would be 25 to 120, or 1 to 4.8. These ratios are much closer to the theoretical figure than the ratios of the previous example (which were 1 in 12 and 1 in 3). If you continue to increase the number of throws, the real-life ratio of fives to throws will continue to get closer and closer to the theoretical ratio of fives to throws.

This law paves the way for a massive expansion in the usefulness of probability. Because it means we don't have to worry about working things out beforehand anymore. You don't have to draw probability space diagrams, or tree diagrams, or any other kind of diagram. And here is why. Imagine a massive supermarket full of shoppers working their way down endless aisles of products towards an array of fifty checkouts at the end manned by fifty disinterested checkout supervisors. Although we can start with a one in fifty chance of being right, there is no way you could accurately predict which checkout one of the shoppers would decide to use. There are too many factors influencing the choice: the last product they bought, the lengths of the queues, superstitions about numbers, aversion to the facial hair on the woman manning checkout 34, and so on and so on and so on.

Fortunately, you don't have to predict. Say you want to know the chances are that a particular shopper will decide to use checkout 29. Then you need to keep track of how many people use this checkout (let us say it is forty-three), and how many people use all the checkouts in total (let us say it is 1500), and you can be fairly sure that the chances of someone using checkout 29 are 43 in 1500, as long as there are no significant changes in the supermarket that might alter the way people choose their checkout. If you wanted to be more accurate, and you had plenty of time on your hands, you could count more people.

Now, this approach to probability is really just another extension of commonsense. People have always been fairly confident that the sun is going to rise in the morning, even before Newton came up with laws to explain why it would. Their argument was simply that, since it had risen every day of their lives so far, and every day of their parents' lives, and presumably every day in recorded history, then the chances were that it would rise the next day, seeing as there was no particular change in the state of the universe that would suggest it might not.

There is an important assumption behind this type of thinking, which is that the universe is an ordered system. If it is not, then there is no basis for the argument above. We cannot make any deductions from past events, because they have no effect on what will happen in the future. We will be in the same position as Bertrand Russell's famous philosophical chicken, which made the mistake of looking forward to a long and leisurely life, because a farmer came and fed it lashings of the highest-quality grain every day. Only on its last day, when it felt the farmer's hands close brutishly around its neck did it realise the mistake it had made. But it never got the chance to report its discovery to its family and friends, which is why chickens cluck so contentedly in their coops, and why they squawk in such astonishment when they are lifted out of their homes by their throats.

However, instead of thinking dark thoughts about the similarity between our position in this world and that of a chicken, let us trust that there is some regular pattern behind our surroundings. If we do this, then the Law of Large Numbers can continue to help us in our daily affairs. Laplace points out helpfully that it should encourage us to always employ happy people, since, if a person has spent a high proportion of his days being cheerful, the likelihood is that he will carry on being cheerful, as long as the circumstances of his life do not change significantly.

The Law of Large Numbers does not just help to give a theoretical basis for things we already thought we knew. It is much more significant than that. For example, everyone was very excited when Newton discovered gravity, and used it to explain why the planets behave as they do. It was a big feather in the cap of Science, and there was a general feeling that others would move on from Newton's work, and that ultimately everything would be explained in terms of such laws.

For example, they hoped to explain the movements of gases using these laws. Gases are simply a collection of particles moving around and colliding, and this situation is not that different from the movements of the planets in space. But they ran into problems, due to the sheer number of collisions and particles present in a body of gas. The situation was too complex for the application of Newton's Laws, which would try and track the path of each individual particle in the gas.

So another approach was needed. And just as it is not possible to predict the movements of one shopper in a supermarket, but perfectly possible to make predictions for a large number of shoppers over a period of time, so it is possible to predict the overall behaviour of a body of gas, without knowing what each individual particle will do. If you assume that each particle is an average particle, with average weight and average velocity, and average everything else, and you apply Newton's Laws to this average particle, the results you discover will hold for the body of gas as a whole.

In other words, you work out what most of the particles are likely to do, and then assume that that is in fact what will happen. This might seem a fairly risky business, but, due to the large numbers of particles involved, it turns out to be the truth. The predictions are as certain as any made by Newton and his Laws.

In fact, when you look closely enough, things that appeared pretty certain and unproblematic suddenly become very unreliable. A table is a pretty average object. You wouldn't expect it to spring any surprises. But you have no real reason for being so confident in this respect. After all, a table is just a bundle of particles held together by electrical forces. At any one time a handful of these particles will gain enough energy to break away from the main body of the table. In other words, for any particle, there is a very tiny chance that it will just up and abandon the main body of the table. But if there is a tiny chance that this will happen for one particle, then there is a miniscule chance that, at one particular time, all the particles in the table will decide to leave the body of the table, and the table itself will vanish.

When viewed in this light, household furniture starts to look suspicious. Fortunately, however, probability applies to the behaviour of the table in the same way as it applied to the behaviour of gases. The table does not disappear, because its behaviour is decided by what the particles it is made of are most likely to do. There is a lot to thank probability for.

10 GAMBLING WITH LIFE INSURANCE

Aside from holding the universe together, the Law of Large Numbers has plenty of other applications, some of them pleasurable and some of them painful. All you need is information, and suddenly you can have a stab at looking into the future.

That is what the Lincolnshire-born Edmund Halley (1656–1742) tried to do in 1693. Beside his discovery of the famous 'Halley's Comet', a notable mathematical career, and the fact that he had the genius to recognize the even greater mathematical genius of Newton (urging him to write the *Principia Mathematica*, and then paying for the costs of publication out of his own pocket), Halley also drew up the first life insurance table. He gave the resulting publication the impressive title of 'An estimate of the degrees of mortality of mankind, drawn from curious tables of the births and funerals at the city of Breslaw; with an attempt to ascertain the price of annuities upon lives'.

Mr Halley felt that it would be very useful if he could work out a rough idea of the chances of a particular person living to a particular age. It took him a while to find a source of information with enough detail on births and deaths to achieve this aim. At the time, the records of London were not suitable, because too often there was no record of the age at which a person died, and because of 'the great and casual Accession of Strangers who die therein.'* It wasn't that Halley didn't like foreigners coming to find work in London, but their habit of dying there meant that there were too many funerals for people

---

* Newman (2001) p1437.

who had not been born there. This state of affairs ruined the records for his purpose. He had to look elsewhere.

What he needed was a sleepy backwater where people liked to keep nice and orderly records. He searched and searched, and finally came across the city of Breslaw in Germany. 'It is very far from the Sea, and as much a Mediterranean Place as can be desired, whence the Confluence of Strangers is but small'. Halley's assumption was that if you were born in Breslaw, then that is probably where you would die, and if you weren't born there, it wasn't the kind of place you would want to visit. It was just what he needed. What is more, the bureaucrats of Breslaw loved keeping records.

Thanks to the conscientious burghers of Breslaw, Halley was able to calculate exactly how many people of each age lived in the city. And from this information he could work out the probabilities of someone of a certain age living for a certain period of time. For example, he explains that a man of forty has a chance of 377/445 of living for a further seven years, since, from the records, of the 445 people aged forty in Breslaw, only 377 made it to their forty-seventh year.

On the basis of such information, life insurance premiums can be calculated. Insurance companies today are no longer relying on the records of Breslaw. They have access to far greater databases of births and deaths, which include a more sophisticated analysis of the population. They have statistics on the effects of gender, social class, lifestyle, and other factors on life expectancy. But their approach to the calculation of their premiums is basically the same as Halley took all those years ago.

The customer asks to insure his life for a certain amount of money. The insurer simply calculates the chances of his losing that money, and comes up with a yearly premium that covers his costs. As long as he has a large enough base of clients, then he can rest easy at night as The Law of Large Numbers ensures that he will come out on top.

For example, say that one of the good burghers of Breslaw came to Mr Halley at the age of forty with thoughts of mortality on his mind, and asked to insure his life over the next year for one hundred crowns. Mr Halley would consult his tables, and argue that there was a chance of 9/445 that he would lose his money, and the man would die within the year. In other words, if 445 such men came and made the same request, he would expect to pay out 900 crowns over the year. To calculate their payment for the privilege of having their life insured, he must divide this cost between the 445 men. So the particular forty-year old man standing in front of his desk, sweating slightly because of the strenuous climb up the stairs, should pay 900 ÷ 445 = 2.02 crowns. However, since Mr Halley also needs to make a living, he might as well say a round 2½ crowns. Then everyone is happy.

It is a very cold business, but Halley was not just in it for the money. He was also able to give some pieces of sound advice, backed up by the statistics in his tables. 'How unjustly we repine at the shortness of our Lives, and think ourselves if we attain not Old Age; whereas it appears hereby, that the one half of those that are born are dead in Seventeen years time... So that instead of murmuring at what we call an untimely Death, we ought with Patience and unconcern to submit to that Dissolution which is the necessary Condition of our perishable Materials, and of our nice and frail Structure and Composition...' He also felt that 'above all things, Celibacy ought to be discouraged'. I shouldn't think he will find many in disagreement with him there.

---

81. There is one piece of cake left. It is homemade and covered with your grandmother's chocolate icing. You want it. Your friend wants it. You decide to toss a coin for it, but the only coin that you have is biased so that the chances of getting a head are 7/10 and the chances of getting a tail are 3/10. How can you toss the coin twice in order to decide fairly who gets the cake?

I don't want to end on a gloomy note, and I can't think of anything more depressing than betting on how long you are going to live. So, here is another application of the Law of Large Numbers that is more cheerful.

Dr Z, whose real name is William Ziemba, and whose 'daytime' job as a mathematician based at the University of British Columbia leaves him enough free time for speculation, recently developed various ways of making money by gambling on horse races. American horse races have a centralised system of betting (called the Tote). You bet on a horse, and if it wins you share the total amount of money placed on the race with all the other lucky punters that picked the right horse. Of course, someone who bets $100 will win twice as much as someone who bets $50. And someone who bets $200 will win ten times as much as someone who bets $20. That is only fair.

Dr Z realised that, although people were reasonably good at picking out a horse that might be a winner, they were not good at working out the odds that the horse might finish in one of the top places. So, he came up with a system that calculated the chances of a horse finishing in one of these places. The Tote displays how much money has been bet on each horse, and the total amount of money bet on the race. Dr Z used this as an estimate of the chances of the horse winning the race. If enough people are betting, then the Law of Large Numbers says that this is a reasonable assumption. So, if $600 dollars had been place on Likely Lad, out of a total pool of $10000 dollars, Dr Z reckoned that the chances of this horse winning were 600/10000.

He then used the rules of theoretical probability to work out the chances of the horse finishing in the top places. Imagine that there is a three-horse race involving 'The Artful Dodger', 'Likely Lad' and 'What A Palaver', and that the probabilities for each of them winning are 5/10, 3/10 and 2/10 respectively. The possible outcomes of the race are shown in the tree diagram on the next page, along with the associated probabilities:

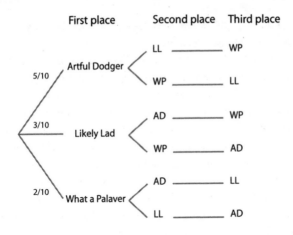

You can see that there are six possible finishing orders. The first one in the tree diagram is The Artful Dodger in first place, Likely Lad in second place and What a Palaver in third. The next one places the horses in this order: The Artful Dodger, What a Palaver, Likely Lad. And so on, and so on.

---

82. You dine with two friends, and the bill comes to £30. You each put down £10 and leave. The waiter realises that he has double-charged you for the fried calamari, and that the bill should have been £25. Since it is not possible to split £5 equally between three, he rushes after you, tells you that there has been a mistake, gives you each a pound, and keeps the remaining £2 for himself. You have now each paid £9, making £27 in all, whilst the waiter has kept hold of £2. £29 is accounted for, but initially you paid £30. What has happened to the extra pound?

The problem comes in working out what the chances are of a particular horse coming second, if another of the horses comes first. This is not an easy problem at all, but the argument goes something like this. You know the probabilities for each horse to win, and they tell you that if the same race was run with the same horses under the same conditions ten times, then you would expect The Artful Dodger to win five times, Likely Lad to win three times, and What a Palaver to win two times.

If The Artful Dodger wins, it is a matter of working out the chances of each of the other two horses of coming second. But from the initial probabilities, you know that Likely Lad wins a race three times for every two times that What a Palaver wins. So you consider this situation to be just another race – although it is, in fact, a race for second place. With the information above, if Likely Lad and What a Palaver raced for second place five times, you would expect Likely Lad to perform the better in three of those races, and What a Palaver to perform the better in two of them. Therefore, if Artful Dodger comes first, the chances of Likely Lad coming second are 3/5 and the chances of What a Palaver coming second are 2/5.

Similarly, if Likely Lad crosses the line first, you can argue that, for every five races that The Artful Dodger wins, What a Palaver wins just two. Therefore, in seven races for second place between the two of them, you expect the Artful Dodger to come out on top five times, and What a Palaver to succeed twice. If Likely Lad comes first, the chances of Artful Dodger coming second are, therefore, 5/7, and the chances of What a Palaver coming second are 2/7. A similar argument can be constructed to find the relevant probabilities if What a Palaver comes first, and now you have all the information that you need in your tree diagram:

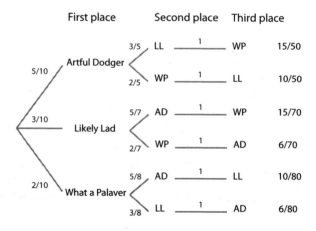

Since this is a three-horse race, if The Artful Dodger comes first, and Likely Lad comes second, then What a Palaver is certain to come third. Therefore, the chances of What a Palaver coming third in this scenario are 1. Similarly, for all the other possible finishing orders, once the first and second places have been decided, it is certain (i.e. probability = 1) which horse will come third.

The probability of each different finishing order is found by multiplying along the branches of the tree diagram. So, for example, the probability of Likely Lad coming first, with The Artful Dodger in second, and What a Palaver in third is given by $3/10 \times 5/7 \times 1 = 15/70$. These probabilities have been included at the end of each of the branches on the tree diagram.

Now, finally, you can help Dr Z with his calculations. If you want to know the chances of What a Palaver finishing in the first two places in this race, you identify which finishing orders satisfy this condition, and then add together their respective probabilities. In this example, there are four orders that you are interested in: AD WP LL, LL WP AD, WP AD LL, and WP LL AD. The respective probabilities for each of these are $10/50$, $6/70$, $10/80$ and $6/80$. You add them together to get $1360/2800$, and this is the probability of What a Palaver coming first or second. Similar calculations can be done to find other relevant probabilities.

---

83. In Rwanda, each panel of judges for the local court contains nine persons of integrity. From these nine people, first a president is elected at random, then two vice-presidents, and then two secretaries. If there are four women on the panel, and if the president is a man, and the vice-presidents are a man and a woman, what is the probability of electing two secretaries of different sexes?

You can imagine how complicated these kinds of calculations became for races with more horses running. To get round this, Dr Z took along a hand-held computer to the racetrack to do the work for him, since he needed his results quickly in order to give him a chance to actually go and place money on a horse when he discovered an opportunity.

Say that, using his computer, he discovered, as you have done, that the chances of What a Palaver coming in the first two places were 1360/2800 (or, as a decimal, 0.486). He then compared this figure with how much money was actually being bet on this possibility with the Tote. If he looked up, and saw that $2800 had been placed in bets on horses finishing in the top two, and only $1000 dollars of this had been placed on Likely Lad, then he reckoned that the public had underestimated Likely Lad's chances (since 1000/2800 is only 0.358 as a decimal), because his calculations showed that $1360 dollars ought to be staked on this possibility. At this point, he rushed to the betting window and backed Lucky Lad to the hilt. But if he looked up and saw that, of $5460 bet on horses finishing in the top two, $3125 had been bet on Likely Lad, then he figured that the public had overestimated the horse's chances (since 3125/5460 is equivalent to 0.572), and stayed well away.

Dr Z's system is not without its flaws. There are things it does not take into account. Certain horses, for some reason, do not like to win, but prefer the comfort and anonymity of second- or third-place. In contrast, there are some jockeys that are only interested in first-place, and so they will let their horse fall out of the leading positions if they can see there is no chance of absolute victory. Variables like this are difficult to take into account. But, even so, over a period of years, Dr Z made a 12% profit on his investment in the horse-racing world. He may well be on to something.

CLOSURE

The minute-hand clicks to the hour mark, and the bell blares. The impatient silence that has settled over the school building is fractured, as hundreds of books are slammed shut, and roughly shoved into bags and rucksacks. Teachers insist on one final ritual, ordering their students to stand quietly behind their desks. Chairs scrape, feet shuffle, and excited voices are stilled for just as long as it takes to win the teacher's approval.

'You can go now...': the sound of release. Ordered ranks disintegrate, and the students surge down the scarred corridors of the school building, filling them until they are bursting with noise. Teachers remain cautiously in their rooms and offices, aware that they have relinquished their control.

Mr Barton looks at his abandoned classroom with its tattered posters of geometrical shapes. Chairs lie upside-down on top of the desks, their legs pointing lifelessly at the ceiling, like the dry bodies of dead insects. He rubs columns of figures and scrawled explanations from the board. He wanders slowly about the room collecting stray balls of paper from the floor, and grimacing at fresh wads of chewing gum. He places his few personal belongings into his briefcase.

Next, he prepares for the bicycle-journey home. He tucks his beige trousers into his socks, fastening luminous bicycle clips around them, he pulls a pale-green anorak over his rumpled suit, and, finally, he fastens the strangely-shaped bulk of a cycling helmet to his head.

A mass of students floods out into the autumn darkness. Small knots of figures stop to discuss the day's events – but break up quickly. Everyone is anxious to get home. Bernadette waits on the pavement in a school uniform unspoiled by the day, each crease still perfectly aligned. She ignores the chaos of her surroundings and begins to mentally sort the day's events in preparation for the detailed questions of her parents.

Charlie is one of the last children to escape. He had to wait outside the deputy headmaster's office to get back his confiscated football, and then he was detained further whilst a teacher explained to him that moonwalking backwards down the corridor is not an appropriate mode of travel for the school environment.

Finally, after making the required gestures of unfelt apology, he has made it outside. The playground is now almost empty, and the wind hustles broken leaves over the faded yellow lines of the basketball court. Teachers are emerging from their special exit and making their way to their cars.

Charlie looks back at his mathematics classroom. He can see the misshapen figure of Mr Barton standing still in the middle of it, like some creature dazzled into immobility by a sudden light. He looks uncertain about what to do next, now that the structure of the school day has vanished.

Charlie stares at this figure for several seconds, confused by a mixture of thoughts. He feels comfort in this glimpse of the smallness of the mathematics classroom, and relief that its obscure rules and regulations have so little influence on the world that lies beyond it. He is fascinated by the oddly-shaped figure in the room, which, as he watches, lumbers into movement and heads towards the door. And, to his surprise, for one brief moment, he feels sympathy for a man who seems so poorly designed for existence in the environment of shops, parks and swimming-pools, in which Charlie himself is so happy.

Mr Barton reaches a waterproofed arm to the light switch, and the window of the room changes instantaneously from yellow to black. Charlie turns around, and runs homewards down the pavement, whilst number-covered pieces of paper fall out of his unclosed backpack and fall silently to the ground behind him.

It is finished. You have come to the end of the road. You can put the memories of your own personal mathematics classroom back into their box, and double-seal it with masking tape. You won't have to look at its contents ever again.

You can walk out the door, head held high, confident that you are the equal of any sums the world might throw at you. You can share out sweets and marbles without batting an eyelid. The division of pizzas no longer holds any problems. Decimals add up. Percentages make sense. Even algebra isn't the torture it used to be.

It is not just your view of the world that has changed; the world will view you differently as well. You don't have to wither under the glare of an intellectual uncle as a result of a ridiculous Monopoly strategy or an incorrect bid, and you can visit the racing track and the casino safe in the knowledge that your friends will marvel at your uncanny knack of winning (just don't make a habit of it). This is all a massive achievement, although I don't advise shouting 'Eureka!' and running naked around the streets in celebration. The Greeks were much more understanding about things like this than we are today.

All in all, you are a better, happier, more rounded individual. There is a more confident tone to your whistle, a jauntier step to your stride, a more authoritative ring to your voice. So, if you see a man standing outside a supermarket asking passers-by mental arithmetic question, don't dive behind the trolley queue and conceal yourself using plastic bags. Don't call the police and have him led away in chains as a public nuisance – although you would probably be within your rights. March up to him, look him straight in the eye, demand that he ask you the most difficult question on the sheet, answer it correctly, kick him in the shins, and steal his pens. It is the right thing to do.

APPENDIX A DIVIDING FRACTIONS

'Turn the second fraction upside down and multiply'. Why on earth should you do that? Won't it make the second fraction feel a bit queasy? And isn't it unethical to order anything to multiply? Surely everything has a choice about when and where it goes about this sort of thing.

Well, after a bit of thinking, it does turn out to be a sensible thing to do. You are making use of a numerical trick to change the division of fractions into a multiplication. You know how to go about multiplying fractions, so as long as you are happy about this trick, everything is just fine.

Take the following division sum: $2/3 \div 3/5$. Remember that a fraction can be thought of as a division sum. So this problem could be written as:

$$\frac{\frac{2}{3}}{\frac{3}{5}}$$

$2/3$ is now the numerator of a new fraction, and $3/5$ is the denominator. (Fractions in which either the numerator, or the denominator, or both, are themselves fractions are called compound fractions.)

Firstly, you can multiply the compound fraction by one, because this won't make any difference to the sum.

So:

$$\frac{\frac{2}{3}}{\frac{3}{5}} \times 1$$

But 1 can be written as a fraction where the numerator is the same as the denominator, since anything divided by itself is 1. So 2/2, 10/10 are the same as 1, and so, more importantly, are things like:

$$\frac{\frac{1}{3}}{\frac{1}{3}} \quad \text{or} \quad \frac{\frac{3}{8}}{\frac{3}{8}}$$

This means that you can rewrite the initial division as the multiplication of two compound fractions without changing its value. For example, you could now write it as:

$$\frac{\frac{2}{3}}{\frac{3}{5}} \times \frac{\frac{3}{8}}{\frac{3}{8}}$$

Now you can multiply the numerators and the denominators according to the rules of multiplying fractions. In this case, the numerators (2/3 × 3/8) come to 6/24, and the denominators (3/5 × 3/8) come to 9/40. So you have transformed the initial sum to a new one, without changing its value. In this case, the new division is:

$$\frac{\frac{6}{24}}{\frac{9}{40}}$$

This is all well and good, but you still have a division of fractions on your hands, and so you are no nearer to solving the problem. What you want is to find a compound fraction that is equivalent to 1, but which has a denominator such that, when you multiply the two compound fractions together, the denominators multiply together to give an answer of 1.

This sounds a bit complicated, but its benefits will become clear over time. For the moment, you just have to find a fraction that multiplies 3/5 to give an answer of 1 – and that fraction is 5/3. 3/5 × 5/3 is equal to 15/15 which is equal to 1. In general, if you multiply any fraction by the fraction that you get by flipping it 'upside down', the answer will always be 1. So, why don't you rewrite the initial division (without changing its value) like this:

$$\dfrac{\dfrac{2}{3}}{\dfrac{3}{5}} \quad \times \quad \dfrac{\dfrac{5}{3}}{\dfrac{5}{3}}$$

If you multiply the numerators of these two compound fractions, you get 2/3 × 5/3, which you can work out according to the normal multiplication rules for fractions. And, if you multiply the denominators, you get 3/5 × 5/3, which you have already worked out to be 1.

At this point you have transformed the initial division (without changing its value) into the following compound fraction:

$$\dfrac{\dfrac{2}{3} \times \dfrac{5}{3}}{1}$$

Now, if you think of it as a division, you are dividing by one, and dividing by one never changes anything. All you have to worry about is working out the top bit. You have transformed the division of 2/3 by 3/5 into the multiplication of 2/3 by 5/3 (or 2/3 by 3/5 'flipped upside-down'). You know how to do this multiplication. $2/3 \times 5/3$ is equal to $10/9$ or $1\ 1/9$. Since, at no point in all this long process have you done anything that will change the value of the sum, this is the answer to the original division

The explanation of this 'trick' is long, but the upshot of it is that you can transform the division of any two fractions into a multiplication in a similar way. You will always be left with the multiplication of the first number in the division by the second number 'flipped upside-down'.

APPENDIX B PUTTING SUDOKU TO BED

I returned home recently after a long absence, expecting to receive a warm welcome from my parents as I walked through the kitchen door. The dog was pleased to see me – jumping up to greet me so that I instantly doubled over in pain – but my parents barely spared me a glance. They were bent intently over a newspaper.

It was Sudoku that was destroying the family dynamics. My parents had started a puzzle categorised as 'Hard', and they had got stuck – just like a pair of inexperienced climbers who have mistakenly started a route graded 'V. Diff.' and found themselves the wrong end of an overhang. They should have stuck to 'Moderate'.

Eventually, I persuaded them to put the problem away, but I could see it was rankling and festering in the back of their minds. Their interest in my recent life was forced – they couldn't wait to get back to the newspaper. I even caught them glancing over my shoulder to where it lay.

I don't want this kind of scenario to become widespread. And so I have made a short study of Sudoku problems, the results of which I wish to share with you, so that you can spend more time feeding the ducks in the park and less time frowning at a collection of numbers and boxes. No puzzle should prevent families from sharing life experiences with one another.

I am sure that what I am going to say is not new, but, as far as I can see, there is a system that can crack any Sudoku puzzle, however difficult, as long as it can be solved purely by logic. At least, I haven't found a puzzle on which the system doesn't work.

Frankly, I find the system long and tedious, but at least that means you can say that you gave up on a puzzle because you lost interest, rather than because you weren't able to do it. It is surprising – and disturbing – how many people look at you with more respect, if you can accidentally leave a completed 'Fiendish' puzzle around the place. The world is such a shallow place.

The first thing to do is to laboriously record in each square what numbers could possibly go in it:

X	Y	3		5			
	5						
		1					
7							
	9						
9							

For example, if you consider the square marked X in the grid above, by looking along the row that it is in, you can discount 3 and 5 as possibilities for it; by looking at the box (or 3 × 3 group) it is in, you can discount 1, 3 and 5 as possibilities for it; and by looking at the column it is in, you can discount 7 and 9 as possibilities for it. Therefore, you can write 2, 4, 6 and 8 as the only available numbers for this particular square.

In the same way, for square Y, 3 and 5 are not possible because they already appear in its row; 1, 3 and 5 are not possible because they already appear in its box; and 5 and 9 are not possible for it because they already appear in its column. The only numbers that remain available are 2, 4, 6, 7 and 8.

After you have noted in every empty square the possible numbers that could appear there, you can fill in any squares where there is only one possibility. Once you have filled in a square, you must look at the row, column and box that it is in, and delete this number from any square where it was a possible option.

★★★

2		157			369	237	2391	589
	23							
57	57							
138								
49								
128								

In the imaginary example above, only some of the squares have been marked with the possible numbers that can go in them. (In reality, all squares will already contain their number, or contain a list of the number that might belong in them.) From the diagram, you can see that the top left-hand square must contain a 2 – it is the only possibility. Once you have filled in a 2 in that square, it is not possible for another 2 to appear in the same row, column or box (from now on abbreviated to R–C–B) as it. Therefore, you must cross out any 2 that is listed as a possibility in any other square in the same R–C–B.

For the example above, the effect is as shown over the page:

2		157			369	2̶37	2̶391	589
	2̶3							
57	57							
138								
49								
1̶28								

Since the top left–hand square now contains a 2, in every square that a 2 is no longer a possibility for that square, the 2s have now been crossed out.

★★★

The third stage is to systematically start checking R–C–B to see if there are any situations where there is only one square in a R–C–B which can contain a particular number.

138	19	167	**4**	2678	369	237	1239	589
			289	**1**	58			
			57	578	57	678 569		
								578
					578			
					345			
								289

In the diagram above, the larger, bold numbers have already been filled in. If you look along the top row, only the top-right-hand square (which is bracketed) has the possibility of having a 5, and so it must contain a 5.

In a similar way, if you look at the top-middle box, only the top-right-hand square (also bracketed) can contain a 3, and so it must contain a 3. Once you have filled in these results, you must, as before, delete any 3s that are listed as possibilities in the R–C–B, which contain the new 3, and you must delete any 5s that are listed as possibilities in the R–C–B which contain the new 5.

The diagram below shows the numbers which now need to be crossed out:

138	19	167	**4**	2678	**3**	237	1239	**5**
			289	**1**	58			
			57	578	57	678	569	
								578
					578			
					845			
								289

★★★

The fourth stage can be combined with the third stage. In the case of rows, you are looking for a situation where a particular number only appears within a particular section in the row. In the case of columns, you are looking for a situation where a particular number only appears within a particular section in the column. In both cases, this means that that number cannot appear anywhere else in the box containing that section.

269	79	1356		**247**				
23	**4**	239	157	579	5789	168	357	1579
58	1679	28		4569				
			1247	**3**	12589			
			569	16	46			
			4789	167	128			
				79				
				8				
				2459				

In the diagram above, if you look at the second row down, you can see that 2 only appears in the left-hand section of it. Although you do not know which square it is in, you know that it is impossible for a 2 to appear elsewhere in the top-left box, and so you can cross out any 2s that are listed as possibilities in other squares in this box.

If you look at the middle column, you can see that 1 only appears in the middle section of it, and therefore it is impossible for a 1 to appear elsewhere in the middle-middle box. You can cross out any 1s that are listed as possibilities in other squares in this box.

~~2~~69	79	1356		247				
23	**4**	239	157	579	5789	168	357	1579
58	1679	~~2~~8		4569				
			~~1~~247	**3**	~~1~~2589			
			569	16	46			
			4789	167	~~1~~28			
				79				
				8				
				2459				

In the diagram on the previous page, you can see that any other 2s in the top-left box and any other 1s in the middle-middle box have been crossed out.

★★★

The same technique applies when you look at different boxes. If you find that within a particular box, a particular number only appears in a particular column, then that number cannot appear elsewhere in that column, and if you find that within a particular box, a particular number only appears in a particular row, then that number cannot appear elsewhere in that row.

269								
14								
8								
237	1467	18						
1346	**5**	37						
23	3478	**9**						
237						**9**	126	2356
1479						135	**8**	124
5	67	348	**9**	346	1278	237	146	237

In the middle-left box (referring to boxes by their row first and then their column), 2 only appears in the left section of the box. You don't know exactly where the 2 goes, but you know it cannot appear anywhere else in the entire of the left-hand column, so you can cross it out in any of the other squares in this column where it is listed as a possibility.

In the bottom-right box, 7 only appears as a possibility in the bottom section, and so it cannot appear anywhere else in the entire of the bottom row, so you can cross it out in any of the other squares in this row where it is listed as a possibility.

2̶69								
14								
8								
237	1467	18						
1346	**5**	37						
23	3478	**9**						
2̶37						**9**	126	2356
1479						135	**8**	124
5	6̶7	348	**9**	346	12̶78	237	146	237

Any other 2s in squares in the left-hand column, and any other 7s in the squares in the bottom row can be crossed out.

★★★

There is one final strategy that helps to speed up the process. As you work through R–C–B, it is worth looking out for two squares in the same R–C–B, which contain the same two possibilities, or three squares which contain the same three possibilities, or four squares which contain the same four possibilities, and so on, and so on. In each of these cases, although it is not possible to say where exactly the numbers go, you know that they must lie in these particular squares in the R–C–B and no others. Therefore, if you come across a situation like this, you can cross out the numbers involved as being possibilities in any other square in the relevant R–C–B.

In the second row down, there are two squares which have the same two possibilities for them: 2 and 3. It is not possible to say which number goes in which square, but it is certain that 2 and 3 cannot be in any other squares in that row.

23	589	1	356	23	4	89	7	5689
	125							
	147							
	147							
	289							
	3					2678	49	4589
	4678					5	39	23678
	5789					39	2349	1

In the second column from the left, there are three squares which have the same three possibilities: 1, 4 and 7. Therefore, these numbers are no longer possibilities for any of the other squares in that column.

In the bottom-right box, there are two squares with the same two possibilities: 3 and 9. Therefore, these numbers are no longer possibilities for any of the squares in that box.

23	589	1	356	23	4	89	7	5689
	125							
	147							
	147							
	289							
	3					2678	49	4589
	4678					5	39	23678
	5789					39	2349	1

In the diagram on the previous page, in the second row down, any other 2s or 3s have been crossed out. In the second column from the left, any other 1s, or 4s, or 7s, have been crossed out. In the bottom–right box any other 3s or 9s have been crossed out.

★★★

In my experience, these techniques should solve any Sudoku puzzle. With harder problems, the going is slow to start off with. After finding the possibilities for every empty square, you are unlikely to find any that have only one possibility. You need to systematically work through row after row, then column after column, and then box after box, using the methods from stages 3 and 4. Gradually, they will help you to fill in some squares, and discount possibilities for other squares, until you get to a point where suddenly it is possible to fill in squares like there is no tomorrow. At this stage, it is probably best to abandon the system and follow your nose – you are likely to solve the Sudoku quicker – but, if you get stuck, you can always return to it.

To show that all that I have said is not just empty boasting, here is how a 'fiendish' Sudoku (puzzle 76 in *The Times Su Doku Book 2* compiled by W. Gould) can be solved. First, it is necessary to work out the possible numbers for each square:

	1	2	3	4	5	6	7	8	9
A	13468	1236	2346	1278	**5**	12379	169	3679	1679
B	134	123	**9**	127	12347	**6**	**8**	37	157
C	1368	**5**	**7**	18	138	139	**2**	**4**	169
D	13567	**8**	356	**9**	127	**4**	56	267	2567
E	**2**	179	45	157	**6**	157	459	789	**3**
F	45679	679	456	**3**	27	**8**	4569	**1**	245679
G	679	**4**	**8**	1267	127	127	**3**	**5**	1269
H	3569	2369	**1**	**4**	238	235	**7**	2689	2689
I	3567	2367	2356	125678	**9**	12357	147	268	12468

From the diagram, you can see that there is no square where there is only one possibility for the number that can go in it. It is necessary to start systematically checking the rows, columns and boxes to see if you can use the techniques mentioned above in stages three and four. You can use the two techniques simultaneously, and if you do, you can defeat this Sudoku as follows:

CHECKING ROWS USING THE METHODS FROM STAGES 3 AND 4:

1. Since 4 can only be in A1 or A3, it cannot be in B1 (example of stage 4)
2. 5 must be in B9, so it cannot be in D9 or F9 (example of stage 3)
3. 8 must be in E8, so it cannot be in H8 or I8 (stage 3)

CHECKING COLUMNS USING THE METHODS FROM STAGES 3 AND 4:

4. 4 must be in B5

CHECKING BOXES USING THE METHODS FROM STAGES 3 AND 4:

5. Since 2 can only be in D5 or F5, it cannot be in G5 or H5 (stage 4)
6. Since 5 can only be in E4 or E6, it cannot be in E3 or E7 (stage 4)
7. E3 must be 4 – and so A3, F3, E7 and F1 cannot be 4
8. E7 must be 9 – and so E2, A7, F7 and F9 cannot be 9

At this stage, you have crossed out several possibilities, but there are no longer any squares where there is just one possibility. You have to start working systematically again, but you can start by looking at areas where several possibilities have been crossed out. In this case, a couple of possibilities were deleted in row A, and so you examine that row using the same techniques as before:

9. A1 must be 4
10. A4 must be 8 – and so C4, I4 and C5 cannot be 8
11. C4 must be 1 – and so B4, E4, G4, I4, C1, C5, C6, C9 and A6 cannot be 1
12. C5 must be 3 – and so C1, C6, H5, and A6 cannot be 3
13. C6 must be 9 – and so C9 and A6 cannot be 9
14. C9 must be 6 – not C1, A9, D9, F9, G9, H9, I9, A7, A8
15. C1 must be 8
16. H5 must be 8 – not H9
17. A7 must be 1 – not A2, A9, I7

Once again, there are no more squares with only one possibility remaining, and so it is necessary to start checking systematically again:

CHECKING ROWS USING THE METHODS FROM STAGES 3 AND 4:

18. I9 is 8

CHECKING COLUMNS USING THE METHODS FROM STAGES 3 AND 4:

19. G9 is 1 – not G5, G6
20. G5 is 7 – not G1, G4, G6, F5, D5, I4, I6
21. G6 is 2 – not G4, H6, I6, A6, I4
22. G4 is 6 – not G1, I4
23. G1 is 9 – not H1, F1, H2
24. F5 is 2 – not F9, D5
25. D5 is 1 – not D1, E6
26. I4 is 5 – not I1, I3, I6, E4, H6
27. H6 is 3 – not H1, H2, I6
28. I6 is 1
29. E4 is 7 – not E2, E6, B4

30. E6 is 5
31. E2 is 1 – not B2
32. B4 is 2 – not B2
33. B2 is 3 – not B1, B8, A2, I2, A3
34. B1 is 1
35. B8 is 7 – not A8, D8, A9
36. A9 is 9 – not A8, H9
37. H9 is 2 – not H2, H8, D9, I8
38. I8 is 6 – not I1, I2, I3, I7, H8, D8
39. I7 is 4 – not F7
40. H8 is 9
41. H2 is 6 – not H1, F1, A2
42. H1 is 5 – not F1, D1
43. D8 is 2
44. D9 is 7 – not D1, F9
45. F9 is 4
46. A2 is 2 – not A3, I2
47. A3 is 6 – not D3, F3
48. A6 is 7
49. A8 is 3
50. I2 is 7 – not I1, F2
51. I1 is 3 – not I3, D1
52. I3 is 2
53. F3 is 5 – not F7, D3
54. F2 is 9
55. D3 is 3
56. D1 is 6 – not F1, D7
57. F1 is 7
58. F7 is 6
59. D7 is 5

As I said, once you pass a critical point, it is suddenly possible to start filling in squares very quickly, but until you get there, you have to be very patient. If you know that you aren't very patient, then you should simply take comfort in the fact that you could do any Sudoku puzzle if you wanted, and go and do something else instead.

APPENDIX C ANSWERS TO PUZZLES

1. Sixty days (the lowest common multiple of three, four and five).
2. Two prisoners share each corner cell.
3.

4. Each man should carry twenty bins. To divide the garbage equally, give the twenty half-full bins to the first man, ten full bins and ten empty bins to the second man, and ten full bins and ten empty bins again to the third man.
5. Ninety-four presents. The number of presents you have left after meeting each of the five cousins are as follows: forty-six, twenty-two, ten, four, one.
6. a) 586.
   b) 3 jackson-fives, 3 high-fives, 4 fives and 3.
7. A = 4, B = 8, C = 9, D = 1, E = 3, F = 6 and G = 5.
8. The first child has eight chocolates, the second child has twelve, the third child had five, and the fourth child has twenty.
9. a) 104 fish are dead at the end of the day.
   b) It is important for your pride that you now score twenty-six points.
10. The cheapest way to make the endless chain is to open up all five links of one piece, and use them to join together the remaining five pieces. If you do this, you will make a saving of

20 cents over the price of a new chain.

11. 215 Pringles are inside you.

12. He was asking for 18,446,744,073,709,551,615 pieces of grain, which is more than the whole world could produce in several years.

13.a) $8 \times 25 = 200$ and $(2 \times 6) + 1 = 13$. $200 + 13 = 213$

    b) $(2 + 3) \times 75 = 375$ and $(9 + 1) \div 5 = 2$. $375 + 2 = 377$.

    c) $6 \times (100 + 7) = 642$. $9 - 4 + 1 = 6$. $642 + 6 = 648$.

    d) $9 + 2 = 11$ and $11 \times (75 + 3) = 858$. $858 + 4 = 862$.

14. One quick way is to pair 100 with 1, then 99 with 2, 98 with 3, 97 with 4, and so on. Each pair sums to 101, and there will be 50 pairs in all (the last one being 50 and 51). So the total of all the numbers is $50 \times 101 = 5050$.

15. You will have written twenty-one ones.

16. 'Twice four and twenty' could refer to the numbers twenty-eight or forty-eight, but since only the first of these can be divided by seven, the poem must mean the former. So, four birds were killed, and twenty-four flew away. The four dead birds are the ones that 'stayed'.

17. 893 people fled.

18. $54 \times 3 = 162$

19.

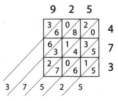

20. If a number is divisible by 8, it is also divisible by 4 and 2. If a number is divisible by 9, it is also divisible by 3. Numbers that are divisible by 8 and 9, are divisible by 2 and 3, and therefore by 6 as well. So you are looking for a number that is divisible by 5, 7, 8 and 9. The Lowest Common Multiple of these numbers is 2520.

21. You form fourteen columns.

22. You can claim fifteen fluffy dice – twelve from the initial forty-eight tokens, and then three more from the bonus twelve tokens.

23. Each villager receives 32 cm of sausage.

24. Call the five litre container A, and the three litre paint-tin B. You fill A and then fill B from it, leaving two litres in A. Then empty B back into the vat and pour the two litres into B. Next, fill A and then fill B from it. It takes one litre of liquid from A to fill B, which leaves four litres in A (as well as traces of yellow paint and lubricating oil).

25. I needed 1100g of beans, 22 onions, 33 tomatoes, 16 ½ tablespoons of brown sugar, 11 tablespoons of mixed spice and 16 ½ teaspoons of salt. Everyone said that they were the best beans that they had ever tasted, and that, if they had been offered turkey, they would have turned it down.

26. The dog should get 460g and the cat should get 115g. You shouldn't really obey the instructions of dream animals, though.

27. There are 126 people who need more fulfillment in their lives.

28. One move. Pick up the third tumbler from the left and pour its contents into the last tumbler.

29. The front of the train will travel the 2km through the tunnel in 45 seconds, but the rear of the train will take a further 4 seconds to emerge from the tunnel. Overall, the train takes 49 seconds to pass through the tunnel.

30. They will plant 480 saplings.

31. In 23 miles of pursuit, you catch up 8 miles. Therefore, it would take you 115 miles to catch up 40 miles. You would have to continue for another 92 miles.

32. 9/24 of the shape is shaded.

33. 100 is the same as 99 9/9.

34. If there are many more squirrels in my garden, then it is perfectly possible that ¼ of the squirrels in my garden is a greater number than ½ the squirrels in your garden.

35. You can first measure 11cm, and draw a mark. Then line the 7cm ruler up so its end is at the mark, and line the 11cm ruler up with the 7cm ruler, so that both rulers start in the same place. The 11cm ruler will project past the mark by a distance of 4cm. You can mark this distance, and it will be 15cm from where you originally started.

36. 41 out of 60 can be thought of as the fraction 41/60, which is

equivalent to 205/300, and 33 out of 50 can be thought of as the fraction 33/50, which is equivalent to 198/300. Therefore, the proportion of French people that thought bad things about the English was greater than the proportion of English people that thought bad things about the French. In other words, there was more love for France in England than there was love for England in France.

37. ½ + 2/5 is 9/10. Therefore the third candidate must get 1/10 of the vote.

38. Overall, during each day-and-night period, the lion will climb 1/6 of a metre. So after 114 day-and-nights, it will have climbed 19 metres, and after 116 day-and-nights it will have climbed 19 2/6 metres. During the 117th day it will reach a point 19 5/6 from the bottom of the pit, but slip back during the night to a height of 19 3/6. During the 118th day it will climb up ½ a metre and reach the top of the pit, escape, and take its revenge on its captors.

39. a)   b)

40. 2 + 4 + 6 + 0.8 = 12.8 and 1 + 3 + 7 + 9/5 = 12.8

41.

Each symbol is one of the numbers with its mirror reflection. 8 is the only number not present. It is also a magic square with magic number fifteen.

42. Your company must pay you £556.50 in petrol expenses.

43. The first bunch of roses costs 64p per rose, and the second costs 69p per rose. You quickly calculate that this is too expensive, and pick your wife some daffodils from the side of the motorway instead.

44. The percentages of people in the room with different characteristics can be represented in a diagram.

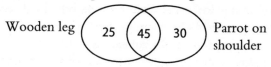

Wooden leg   25   45   30   Parrot on shoulder

Since the percentages must add up to 100, 45% of the people in the room have a wooden leg and a parrot.

45. Last year's membership fee was $1208.33 (to the nearest cent)

46. 56% to the nearest whole number.

47. The area of the rectangle is given by 10 × 4. If the length of the rectangle is increased by 20%, then the new length will be 12. In order for the area to remain the same, the new width must be 3 1/3, which is a 16 2/3% decrease on the original width.

48. It does not matter. The three discounts have the effect of multiplying the initial price by 0.9, 0.85, and 0.8 respectively. It makes no difference in which order you do these multiplications.

49. 82cm to the nearest centimetre.

50. 30 % of students got both questions correct, and so nine students are equivalent to 30 % of the class. Therefore, the whole class consists of thirty students.

51. Members will have to pay $1446.38 (to the nearest cent)

52. If you call x my initial weight, and a the amount of weight I puts on by eating green bananas, then the information given leads to two equations. The first is $x + a + 10 = 1.1x$, and the second is $x + 2a + 10 = 1.11x$. If you solve these two equations simultaneously, you find that my initial weight was 111 1/9 pounds.

53. $2000 + 20m$

54. If the number you think of is x, then, as you work through the calculations, in terms of x, you have: $x + 4$, then $2 \times (x + 4)$ which is $2x + 8$, $2x$, and x. So whatever number you start off with, you end up with the same one.

55. The length of the head is 6 cm.

56. If the time past is x hours, then the time left in the day will be $12 - x$ hours. So $2x = 12 - x$, and the solution is that four hours have passed. Or you can just work it out using common sense.

57. Let x be the number of days he catches a fish, then $(30 - x)$ is the number of days he does not catch a fish. Then, the total amount of money he makes will be $3x$, and the total amount of money he pays out will be $2 \times (30 - x)$. Therefore, $3x = 2 \times (30 - x)$, and this equation has the solution $x = 12$.

58. The width of the rectangle is 6cm.

59. Call the number of frogs, x, and the number of princes y.
If there are 35 heads, then $x + y = 35$. If there are ninety-four feet, then $4x + 2y = 94$. The solution to this pair of simultaneous equations is $x = 12$ and $y = 23$.

60. Let the share of the son you knew about be x, and the share of the son you did not know about be y. Since £10 000 are shared between them: $x + y = 10\ 000$. Since a fifth of the first son's share is £1100 more than a quarter of the second son's share: $1/5\ x = \frac{1}{4}\ y + 1100$. On solving these two simultaneous equations, you find that $x = 8000$ and $y = 2000$.

61. If each friend pays 8 dollars, and there is a surplus of three dollars, then the actual price is $8x - 3$ dollars. If each friend pays seven dollars and they are four dollars short, then the actual price is $7x + 4$ dollars. Therefore $8x - 3 = 7x + 4$, which gives the solution $x = 7$.

62. If you say x is the numbers of bottles of the first perfume that you put in the mixture, and y is the number of bottles of the second perfume that you put in the mixture. Then the total cost of perfume in the mixture will be $10x + 4y$, and there will be $x + y$ bottles of perfume in the mixture. Therefore the cost of the mixture per bottle is given by $\frac{10x + 4y}{x + y}$ and this must be equal to six pounds.
Rearranging this equation you find that $2x = y$, and therefore the mixture must contain twice as much of the second perfume as the first. In other words a mixture of the two perfumes in a ratio of 1:2 should cost six pounds.

63.

F	B	J
C	A	E
G	D	H

Square A had sides of length x, and so has an area of x^2. Rectangles B, C, D and E each have sides of length x and 2½ units, and therefore they each have an area of $2\frac{1}{2}x$. The total area of B, C, D, and E combined is 10. According to the initial problem, $x^2 + 10x = 39$, and so the combined area of square A, and rectangles B, C, D and E is 39 units. Each of the little squares F, G, H, and J have sides of length 2½, and therefore they each have an area of 6¼. The total area of squares F, G, H and J is 25. The total combined area of the large square (A), the rectangles (B, C, D, and E) and the little squares (F, G, H and J) is $39 + 25 = 64$. But this total area is a square. If it is a square it must have a side of 8 units. One side of the square is equivalent to 2½ + x + 2½, and so $x = 3$ units.

64. You need to take out six earrings to be sure of having a pair.

65. Bert always moves forward three spaces each go. Ernie has an equal chance of moving forward any number of spaces from one to six inclusive. He therefore moves forward an average of 3.5 spaces per go, and is more likely to win.

66. The chances of winning a prize are 3/120.

67. The chances of not getting a sausage are 89/138

68. If the chance that LL will win is 1 then the chance that HH will win is 2 and for RR it is 4. Therefore, the chances of RR winning are 4/7, which means that the chances of him not winning are 3/7.

69. All the time in the world, because the ladder will rise with the ship, and the ship will rise with the tide. Besides, you can hear the ticking of the clockwork crocodile, and Captain Hook looks nervous.

70. The chances of selecting a female administrator are 36/350.

71. If the players play for another point, and the first player wins, then they are in the first situation that we looked at, in which the first player is entitled to 56 pistoles. If the first player loses, then each player will have one point, and they can both claim 32 pistoles. Therefore the first player can claim the 32 pistoles as his. He is as likely as not to win 24 more (to make 56), and so is justified in pocketing 12 of these. All in all, the first player receives 44 pistoles, and the second player receives 20 pistoles.

72. The only way the hunter could have travelled south, then east, and then north to return home is if his house is at the North Pole. So, the bear must be a polar bear – and therefore white…

73. Meetings will continue to take place until no more killings can take place. You do not have to worry about your meetings with pacifists, since they are not going to harm you in any way. So, you just need to concentrate on the killers. Now, each time that two killers meet they both die, and so their numbers will be gradually reduced in twos. Since there is an odd number of them, this means that eventually there will only be one left, and he will run amok, because there is nothing to stop him. Even if you are very fortunate, and avoid the one hundred killers who have perished, you must eventually run into this one, and, whether you are a pacifist or a killer, the result will be the same: game over. Your chances of survival are zero, so you might as well take him with you, if only to save some of the pacifists.

74. The chance of getting on a deluxe coach is ½, because during the day there are the same number of deluxe-coach departures, as there are minibus departures.

75. Twelve people are either men, have grey hair, or are men with grey hair, and so the probability is 12/18.

76. The chances of picking two black beans are 42/90 and the chances of picking two red beans are 6/90, so the chances of picking two beans of the same colour are 48/90.

77. a) You expect roughly four of each number amongst the dice, and so an opening bid of '8 somethings' (but not '8 ones') is a good call.

b) You expect roughly three or four of each number amongst the dice. So you reckon it is likely that there are six threes, especially as you have two amongst your own dice. It is probably best to raise the bidding to 'six somethings' or even 'seven somethings' (excluding 'six ones' and 'seven ones').

c) You expect roughly three of each number. 'seven fives' is only slightly more than this expectation. It is probably good to take into account your own dice. If you exclude your dice, there are fifteen dice around the table, amongst which there needs to be 'six fives' to make your neighbour's call correct (as you have one five amongst your dice). Amongst fifteen dice, you would expect there to be one, two or three of each number, and so again it is difficult to say whether to call your neighbour or bid higher. You will have to go on gut instinct, although a reasonably safe call would be 'six sixes'. This puts the next problem in exactly the same situation as you are now.

78. The probability that a student likes at least one of the subjects is 58/60

79. Given the information that the gun contains 3 bullets in a row, there are six possibilities for how the bullets are arranged in a circle (B = bullet, E = empty):

1. BBBEEE	2. EBBBEE	3. EEBBBE
4. EEEBBB	5. BEEEBB	6. BBEEEB

(5 and 6 count because the bullets are in a circle. If you arrange these orders in a circle, the three bullets will lie next to each other. The first player dies if the bullets are in arrangements 1, 3, 5, and 6. The second player dies if the bullets are in arrangements 2 and 4. I think you want to go second. Your chances of survival are then 4/6.

80. There are only two possibilities: his daughter has a younger brother, or his daughter has a younger sister. Therefore, the chance of him having two daughters is ½.

81. Here is a tree diagram, to show the possibilities after flipping the coin twice:

From the diagram, you can see that the chances of getting HT are the same as the chances of getting TH (7/10 × 3/10), whilst the chances of getting HH or TT are different. To use the coin fairly, you toss it twice. One of you calls heads, and the other calls tails. If you get HH or TT, you discard the result and start again, but if you get HT or TH you take whatever was the result of the first toss (i.e. HT is equivalent to 'heads' and TH is equivalent to 'tails'), since both of these possibilities are equally likely.

82. It is not true that the money should add up to £30. The diners have spent £27, of which £25 was for the meal, and £2 was for the waiter.

83. Now that the presidents and vice-presidents have been elected, there are only three women and three men to choose from. There are two possibilities that satisfy the conditions: the first secretary is male and the second secretary is female (probability: $3/6 \times 3/5$, or $9/30$) or the first secretary is female and the second secretary is male (probability: $3/6 \times 3/5$, or $9/30$). Therefore, the total probability that the secretaries are of different sexes and can indulge in a clandestine workplace relationship is $18/30$.

PUZZLE SOURCES

All puzzles composed or adapted for this book by Lawrence Potter, with the exceptions of puzzles number 3, 10, 16, 20, 28 and 33, which can also be found in *The Penguin Book of Curious and Interesting Puzzles*.

BIBLIOGRAPHY

Books

Adams, M. *The Morley Adams Puzzle Book*, Faber 1939
Always, J. *Puzzles to Puzzle You*, Tandem Books 1965
Bacon, F. *The Essays*, Penguin 1995
Bell, E. T. *Men of Mathematics*, Penguin 1965
Carroll, L. *Pillow Problems and A Tangled Tale*, Dover 1958
Carroll, L. *The Unknown Lewis Carroll*, (ed. Collingwood, S. D.)
      Dover 1958
Descartes, R. *Philosophical Essays and Correspondence*, Hackett
      Publishing 2000.
Descartes, R. *Key Philosophical Writings*, Wordsworth Editions 1997.
Hawkind, T. L. *Jean d'Alembert: Science and the Enlightenment*,
      Gordon and Breach Science Publishers 1990
Eves, H. *Introduction to the History of Mathematics*, Holt, Rinehart,
      and Winston 1976

Fauvel, J. and Gray, J (eds.) *History of Mathematics*, Palgrave
    Macmillan 1987
Flegg, G., Hay, C. & Moss, B. (eds.) *Nicholas Chuquet: Renaissance
    Mathematician*, Dordrecht 1985
Gamow, G. and Stern, M. *Puzzle Math*, Macmillan 1958
Gardner, M. *Mathematical Circus*, Penguin 1981
Geijsbeek, J.B. *Ancient Double Entry Bookkeeping: Lucas Pacioli's
    Treatise*, 1914.
Greek Anthology, The Vol. 5 of the Loeb Classical Library,
    Heinemann 1941
Gould, W. *The Times Su Doku 2*, HarperCollins 2005
Gow, J. *A Short History of Greek Mathematics*, American
    Mathematical Society 1968
Hausch, D., Lo, V. & Ziemba, W. (eds.) *Efficiency of Racetrack
    Betting Markets*, Academic Press, 1995
Ifrah, G. *The Universal History of Numbers*, The Harvill Press, 2000
Jackson, J. *Rational Amusements for Winter Evenings*, Barry and Son
    1821
Joseph, G. G. *The Crest of the Peacock*, Penguin Books 2000
Katz, V. *A History of Mathematics*, Longman 1998
Kline, M. *Mathematics in Western Culture*, Penguin 1990
Kordemsky, B. A. *The Moscow Puzzles*, Penguin 1972
Locke, J. *An Essay Concerning Human Understanding*, Penguin 1997
Loyd, S. *Mathematical Puzzles of Sam Loyd*, (ed. Gardner, M.)
    Dover 1959
Loyd, S. *More Mathematical Puzzles of Sam Loyd*, (ed. Gardner, M.)
    Dover 1960
Loyd (Jnr.), S. *Sam Loyd and his Puzzles*, Barse & Co., New York
    1928
Mcleish, J. *Number: From Ancient Civilisations to the Computer*,
    Flamingo, 1991
Midonick, H. *The Treasury of Mathematics*, Vol.1, Penguin, 1965
Midonick, H. *The Treasury of Mathematics*, Vol.2, Penguin, 1968
Morris, I. *The Ivan Morris Puzzle Book*, Penguin 1970
Morris, I. *Foul Play and Other Puzzles*, The Bodley Head 1972
Newman, J. *The World of Mathematics*, Simon and Schuster 2001

Paulos, J. A. *Innumeracy: Mathematical Illiteracy and its Consequences*, Hill & Wang 2001

Peirce, C. S. *Philosophical Writings of Peirce*, Courier Dover Publications 1955

Phillips, H. *Brush Up Your Wits*, Dent 1936

Phillips, H. *Question Time*, Dent 1937

Phillips, H. *Problems Omnibus*, Arco Publications 1960

Plato, *Laches, Protagoras, Meno & Euthydemus*, Harvard University Press 1924

Rayner, D. *Complete Mathematics*, Oxford University Press 1990

Recorde, R. *The Ground of Artes*, London: Reynold Wolff 1552

Robins G. & Shute C. (eds.) *The Rhind Mathematical Papyrus*, British Museum Publications 1987

Rouse-Ball, W. W. *A Short Account of the History of Mathematics*, Dover Publications, 1960

Sadler, A.J. and Thorning D.W.S. *Understanding Pure Mathematics*, Oxford University Press 1987

Smith, D. E. *History of Mathematics*, Dover Publications, 1958

Trachtenberg, J., Cutler, A (trans.) & McShane, R. (trans.), *Trachetenburg Speed System of Basic Mathematics*, Souvenir Press 1989

Victor, S.K. (ed. & trans.) *Practical geometry in the High Middle Ages: 'Artis cuiuslibet consummatio' and the 'Pratike de geometrie'* American Philosophical Society, 1979.

Wells, D. *The Penguin Book of Curious and Interesting Puzzles*, Penguin Books, 1992

Websites

www.publications.parliament.uk
www.en.wikipedia.org
www.hse.gov.uk
www.evidence-based-medicine.co.uk
www.nsc.org